AF882653

SOTHIS PRESS

Sothis Press aims to completely reimagine the public domain book with an emphasis on universal access to captivatingly designed print editions. We believe that technology will enable every single book ever published to be perpetually accessible in beautiful print form. Sothis is beauty in ubiquity.

www.sothis.press
1 Waterhouse Square, London, ECIN 2ST

Specimens in Eccentric Circular Turning

With Practical Instructions for Producing Corresponding Pieces in That Art

John Holt Ibbetson

SOTHIS PRESS

ISBN 978-1-0368-0679-8

Specimens in Eccentric Circular Turning
is in the public domain.

This edition is published by
Sothis Press © 2023. The company
is registered in England and Wales.
The Company Number is 14900389.

THE SLIDE REST.

At the bottom of Plate 1, Fig. 2, a perspective view is given of the Slide Rest I have always used, never having possessed any other: it is very simple and compact in its construction; and, as far as regards *Eccentric Turning*, is a complete and effective instrument. It is of the construction of Messrs. Holtzapffel, of whom I purchased it many years ago, and by their permission I have introduced the engraving and give a description of it. In the course, however, of the long period I have possessed it, they have made many improvements and additions to their Slide Rests, as might reasonably be expected, and have greatly amplified their power of action.

The stand of the Slide Rest (shewn at A. A.) fastens to the Lathe Bed by a screw, in the usual manner. B. B. is attached to this stand by a stem passing into the projecting boss, and by the screw *a.* in the side of the boss, the stem is clamped fast, so that B. B. shall stand at the required height from the Lathe Bed and at any angular position with respect to the Mandril. C. C. slides from end to end on B. B. and is moved and held fast in its required place by the Slide Rest screw (not seen in the engraving), which is moved by its handle *b.* D. D. slides, backwards and forwards, in the piece C. C. at right angles with B. B. The tool to be worked with, fits into the socket *e. e.* and the square-headed screw at the top of *e. e.* clamps and holds the tool fast. At the end of the slide D. D. there are two adjusting screws *d. d* which regulate the depth and limit the cut of the tool.

The cutting edges of the tools are

ground up to various shapes—angular, square, and rounding, &c. according to the taste and fancy of the workman. In turning down the surface of the ivory or wood that is to be ornamented, I generally make use of a tool with a rounding edge, which I sharpen on a stone by hand; but to grind and sharpen the angular and square, or flat cutting tools, I use a tool made by Messrs. Holtzapffel, by which any desired angle is obtained to a certainty. This contrivance for setting the angular tools is highly convenient, if not absolutely necessary, as it enables the workman to sharpen the tool as often as he pleases, during the progress of a piece of Turning, which, without it, or some contrivance of the sort, it would not be prudent in him to attempt, for fear of altering the angle of the tool, and thereby spoil the work.

THE CHUCK.

I do not wish it to be supposed that I take to myself the credit of being the inventor of *Compound Eccentric Chucks.* Instruments, constructed on various plans, have appeared before the Public under that name—there is a description of one in the Manuel du Tourneur—and Messrs. Holtzapffel, and other manufacturers, have constructed them for a length of period. The one which I describe in the present Edition, happens however to differ essentially, in its construction, from those which have hitherto become known; and I offer it to attention as possessing far greater powers. It is the result of the perseverance and labour of a self-taught Amateur Mechanic and

Turner, having been contrived and constructed by myself many years ago, from the raw materials of brass and steel, with my own hands; and I present it as an instrument to which I have never given publicity, excepting having shewn it and explained it to two or three friends. Had I not proved able to make it myself, my plan of its construction would never have been realized; and I, certainly, should not have had the means of producing the Compound Eccentric work which is exhibited in this Edition.

Six-and-thirty years ago I purchased a Lathe and commenced Turning as an amusement; and I very soon saw that the Art of Turning might be greatly extended by various little contrivances to be appended to the Lathe and Slide Rest, and by the *construction of Chucks:* but I found, at the same time, that one of the greatest difficulties that opposed the pursuits of the Amateur Mechanic, was his not knowing how or where to get his projects carried into practical effect. I invariably found that the professional Mechanic, in the habit of constructing Machinery, considered it too great an interference with his own business to trouble himself with the projects of an Amateur: in fact, I saw plainly, that, to carry my plan of an Eccentric Chuck into practice, without being cramped or foiled in my views, it was absolutely necessary for me to become my own workman; and I, therefore, took resolution and resorted to my own exertions and actual manual labour. I had, moreover, a strong motive for going into the labour of making the Chuck myself, in a wish to keep the mechanism of it secret; and this I full well knew I should have no chance of doing, if the Chuck were manufactured by any other hands than my own: I, in fact, constructed the Chuck in aid of a plan I had conceived for rendering Bank Notes more difficult of imitation.

I most cordially wish that every Amateur Turner would make known, where there is no object for secrecy, in some way or other, whatever might occur to him as being a novelty in the Art; it would prove a source of amusement to himself and of information to others. If he only gave specimens of the result of his labours, without communicating the particulars of his machinery, he would give a great stimulus to exertion in the Art of Ornamental Turning. Besides, there are facilities now open to the Amateur, for carrying his projects into practical effect, which did not formerly exist. There is no longer any occasion for him to be his own workman (unless secrecy be desired)—he can readily get any thing executed for him, let it be ever so extended in its mechanism, or of ever so trifling a nature. If he wants a casting in brass or metal, or a piece of steel forged, or a mere screw made, he is able, without any sort of difficulty, to get it done at Messrs. Holtzapffel's manufactory. This is, unquestionably, a very great practical convenience.—I find it so, and, now, continually avail myself of it.

My Compound Eccentric Chuck, as I now use it, and as it is represented by the Copper Plate Engravings inserted against the title page, has been altered to its present state since I executed the Specimens which were exhibited in the *First Edition*. The alterations I have introduced, considerably simplify its construction, and, at the same time, give it a power of adjustment and of producing some patterns which it did not before possess, and which I had never contemplated as coming within the range of the capabilities of any Eccentric Chuck, however extended its mechanism might be. I, however, profess to give a full description of my Chuck, as I used it in executing *all* the Specimens; and therefore, before I enter into a general description of it, *in its present state,* I will explain the particulars of the alterations I allude to.

On referring to Plate Fig. 1, the two circular moving pieces are seen with their appertaining circles divided into 96, and toothed wheels attached, each cut into 96 teeth, to be acted upon by a click. When the Specimens which appeared in the First Edition were executed, these two circular moving pieces were divided differently—the edge of the one next the Lathe head (marked L. Plate 1.) was *racked on its edge,* so as to be worked by a tangent screw, into 288 divisions,—whilst the edge of the other circular moving piece which is attached to the second Bed Plate was cut into 96 teeth and acted upon by a click. The tangent screw which acted on the racked edge, then in the place of L, was fixed in a frame, in the same manner as the screw *d. d.* in Fig. 1. The frame was moveable at one end as a centre, and at the other end there was a catch which kept the tangent screw fast to the racked edge of the circle, or detached from that edge, at pleasure. The tangent screw was, therefore, susceptible of being employed as a click; at the same time, by turning the screw round, the circle could be stopped at every possible point; and a ring of brass, fixed on the head of the tangent screw, was decimally divided into a hundred parts, so that the power of division was very great. The tangent screw acting on a racked edge is by far the most complete of the two methods adverted to; but there are practical difficulties and mechanical labour attending its execution, which induced me to adopt the *click and toothed wheel* system—and I also found that the number 288, into which the circle was racked, did not accord, in point of angular adjustment, with the number 96, which was the number of teeth in the toothed wheel. The alteration I mentioned consisted, therefore, in making the two circular moving pieces alike in all respects, with a division of the edge into 96 teeth, to be worked by a click. With the machinery I possess, I find, however, no difficulty in racking the edge of a circle into *any number* that is *divisible by some other number:* I lately racked one into Two hundred and seventy-two divisions, which is rather a curious number, being divisible by 136, 68, 34, and 17.

I will now enter into some preliminary explanation of a few of the results which flow from the combined adjustments of the two circular moving pieces I have just adverted to, and of which the additional Specimens will furnish some practical examples. I consider them extremely curious; and I think I may venture to pronounce them, without fear of contradiction, as being completely novel in the Art of Eccentric Circular Turning.

The engraving, Fig. 5, gives a plan of the Bed Plate and Slide Plate of the Chuck; and the circular moving pieces, just mentioned, are affixed to the Slide Plates G. G. in the centre, as represented at G. G, where there are two circular lines described. By means of the Slide Plates, therefore, the two circular moving pieces may be placed eccentric to each other, and to the Mandril's motion. Now, suppose it be required to place 96 circles or dots,* consecutively, in the curve of an Ellipsis, the transverse diameter of which shall be three inches, and the conjugate diameter two inches. The method by which it is to be aecomplished with the Compound Eccentric Chuck is as follows:—Adjust and fix the describing point, or cutting tool, so as to cause its point to coincide with the centre of the Mandril's motion, if it be in-

* The diameter of the circles is determined by the position of the describing point, or cutting tool which is fixed in the Slide Rest. If the point or tool be adjusted to coincide with the centre of the Mandril's motion it will describe a dot, if it be adjusted to the left of the centre of the Mandril it will describe a circle, the radius of which will be equal to the distance between the point of the tool and centre of the Mandril.

tended to describe a dot, or at a distance from the centre of the Mandril equal to the radius of the required circle. Then move the Slide Plate next the Lathe head (G. G. Fig. 5.) by turning the screw (*d. d.* Fig. 1,) $\frac{1}{4}$ of an inch from its central position, and the other Slide Plate $1\frac{1}{4}$ inch. Set the circular moving pieces so that the index points to division 96 on both.—Turn the Lathe round, and cause the fixed tool to describe the dot or circle. The circular moving piece, next the Lathe head, is then to be moved two divisions, making the index point to division 2; and the other circular moving piece is to be moved one division, in the contrary direction, making its index point to division 95. Another circle or dot is then to be described. The first circular movement is again to be moved two divisions, which will place the index at division 4. The other circular movement is to be moved one division, in the contrary direction, which will bring its index to point to division 94, and describe a dot or point as before. The circular movements are to be moved and adjusted, in this way, until the figure be completed: the circular movement next the Lathe head is to be moved two divisions at a time, in its direction, whilst the other circular movement is to be moved one division each time, in the contrary direction.

The number of beautiful designs which may be obtained by combining, on this principle, two circular adjusting movements, is inconceivable. Consecutive circles, &c. may be arranged, not ouly in elliptical curves, but in the shape of hearts—in straight lines—in triangles—in squares—in polygons, and in both inward and outward looped figures. In Elliptical figures the ratio of adjustment, between the two circular movements, is as two to one, in contrary directions; that is, if the one be moved two divisions each time, the other must be moved one division each time, in the contrary direction. If one circular movement be moved four divisions each time, and the other two divisions in the contrary direction, the Elliptical pattern will consist of 48 consecutive circles; if one of them be moved eight divisions and the other four, the number of consecutive circles, in the pattern, will be 24; and thus the number of the consecutive circles can be always regulated. With this same ratio of adjustment of the Slide Plates and of the circular movements, the latter being in the same direction, instead of being in contrary directions, the pattern formed will consist of two loops inwards. If the ratio of adjustment, between the two circnlar movements, be equal (that is, if they each be moved the same number of divisions at every adjustment) and the movements are made in the *same* direction, the pattern will be heart-shaped.

In all figures where the number of angles or loops are more than *two*, the loops will invariably turn outwards when the circular movements are moved in contrary directions, and inwards when they are moved in the same direction.

Triangles are obtained when the ratio of the circular adjustment is as three to one: and, Squares are obtained by adjusting the circular movements in the ratio of four to one. Figures of six loops are obtained when the ratio of the adjustment of the circular movements is as six to one; and so on for all others. In fact, the adjustment of two circles, in the manner I have described, arranges the circles or dots which compose the pattern, in the curve of an *Epicycloid;* and, in conformity with the properties of that curve, the number of loops inevitably depends on the ratio of velocity between the two circular motions. The circles or dots so arranged in the curve of an Epicycloid, will gradually approach each other in one or more parts of the pattern, and recede from each other in other parts; but this inequality of arrangement must not be looked upon as a defect in the Machinery, as, in reality, it results from natural causes inherent in Epicycloidal motion. A compensating movement might be introduced into the Chuck that would correct this varying principle—I have applied it to other apparatus, and, at one time, intended to add it to the Compound Eccentric Chuck, but I have never donc so.

THE ENGRAVINGS.

I will now give a description of the mechanical contrivance of the Chuck which I have denominated "Ibbetson's Improved Compound Eccentric Chuck," and in the course of explaining the copper-plate Engravings of it, inserted in front of the title-page, I shall particularly take notice of those parts which I consider to be new, and that give results which have not hitherto been looked upon as coming within the range of Eccentric Turning.

I would not have it supposed to be my intention to offer instruction to the professional Mechanic, on matters which he may, probably, be already fully acquainted with. I consider that I am addressing the Amateur who may know little or nothing about the mechanism of an Eccentric Chuck; and, therefore, if I am more diffuse in my description than by some may be thought necessary, and enumerate things that have been, over and over again, employed in the construction of Machinery, I will still hope to stand excused; and I beg to be perfectly understood as not taking credit for invention, in any instance, beyond what I shall specially lay claim to.

Figure I gives a side view of the Chuck in its real dimensions, and as it appears when screwed on the Lathe.

Figure 5 gives a Plan of the Bed Plates and Slide Plates of the Chuck; and,

Figure 3 gives a Section shewing the internal construction of the circular movements of the Chuck.

In figure 1 the piece A. B. B. is the Bed Plate of the Chuck next the Lathe-head, and, with the projecting piece D, is formed out of one casting of brass or metal. This Bed Plate screws on the nose of the mandril at its projecting part A. and carries the whole of the remaining parts of the Chuck. The projecting piece *f.* at the end of the Bed Plate opposite to D. is kept to its place by steady pins, and is attached to the Bed Plate by two screws, one of which is seen just under *f.**

The Cut hereunder inserted is a section of the piece *f.* as it would appear across the Bed Plate.

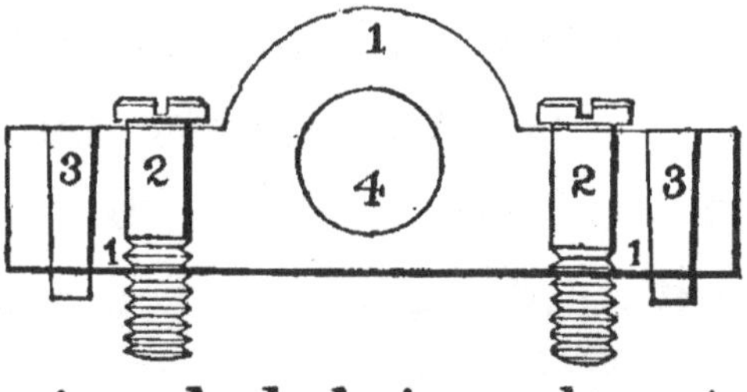

This piece 1. 1. 1. is made out of brass, separate from the Bed Plate, but attached to it by the two screws 2. 2. and kept

* It is to be understood that what is said with regard to this Bed Plate and its appendages, *next the Lathe-head,* applies also to the other Bed Plate and its appendages.

firm to its place by the steady pins 3 3. These steady pins can only enter into the Bed Plate a very little way, as shewn in the cut, on account of the screw *d. d*, one end of which passes through the hole at 4. This piece *f.* (shewn at *f.* in fig. 1.) must be fixed in its place on the Bed Plate before the hole at 4. is drilled; which hole, and also the screw hole through the piece D, must be both drilled at the same time, in the manner hereinafter explained.

In figure 5, B. B. gives the outline of the Bed Plate, at one end of which is seen a portion of the screw *d. d.* fig. 1. G. G. figure 1. gives the outline of the Slide Plate, which it will be perceived is not so long as the Bed Plate, thereby giving more scope for eccentiic adjustment. In fig. 5, E. E. is one steel arm and F. F. the other, between which the Slide Plate traverses. The two circular lines, in the centre, shew the hole through which the screw N. (fig. 3) passes; and the large circle represents the outline of the screw nut *k. k.* fig. 3 L. fig. 5, is a portion of the toothed wheel which forms part of the circular movement, and 5. 4. is the click which stops the toothed wheel at any required division of the circle. In the Bed Plate B. B. (see figures 1 and 5, and Boss A. fig. 1.) there is a long opening, *c.* c. fig. 5, represented by dots, and by dots at Boss A. fig. 1, which gives passage to the plug and screw ring through which the screw *d. d.* acts, see fig. 4. E. E. and F. F. fig. 5, are steel arms with angular edges, which take into the dovetailed edges of the Slide Plate G. G. The steel arm E. E. is retained in its position by two steady pins, represented by dots, and screwed fast to the Bed Plate by two screws. The steel arm F. F. is kept fast to the Bed Plate by two screws, and the holes in it, through which the screws pass, are elongated (as seen by dots), in order to meet the action of the clamp screws *i. i.* The Slide Plate G. G. traverses, parallel to the Bed Plate, between the two steel arms E. E. F. F.; and the screws *i. i.* which screw into the edge of the Bed Plate, are adjusting screws to press the steel arm into the grooved edge of the Slide Plate, so as to cause the Slide Plate to traverse with the proper degree of tightness. The screw *d. d.* (seen in fig. 1.) is the Slide Plate screw, and by the action of it the circular movements, which the Slide Plates carry, (hereafter described) become fixed at any required position eccentric to the mandril. This screw is retained in its place in a different manner to any I have usually seen employed in similar cases, and is so secured that nothing like what is called *loss of time* can possibly occur; and in every application of a screw, fitted as this is, I have found its action most efficient. One end of this screw has a countersunk conical hole drilled in it, to receive the point of the screw which screws *tightly* through the piece D. The other end passes through, and correctly fits, a round hole in the piece *f*. (see 4. in the cut in page 24,) and the screw at D. presses the shoulder of the screw *d. d.* into close contact with the side face of the piece *f*. The part of the end of the screw which passes beyond the piece *f*. is left with a small shoulder flush with the face of the piece. A ring of brass (the ring seen at *f*. with division lines) fits up to this shoulder, and is kept firm to it by a milled-head screw; and the end of the screw, beyond the screw nut, is squared, in order to its being moved round by a key. I divide this brass ring into one hundred, of which, four grand divisions are distinguished by 0, 1, 2, 3; and this ring is made to move round against the said shoulder, in order that 0. may, under any circumstance that may arise, be adjusted to the line serving as an index on the top of *f*. The piece *f*. is also necessarily detachable from the Bed Plate, to make way for the screw *d. d.* being placed in its proper situation for action.

The Click seen at 4. 5. figure 5. is, I consider, an entirely new contrivance of my own invention, and I call it *A double acting Micrometer Click.* This double acting micrometer Click, has never been, however, actually applied to my Compound Eccentric Chuck; but having used it in other machinery, and found it to be very efficient in its action, I resolved to give the present edition the advantage of its publication. The drawing of it in fig. 5 was taken, for the engraver, from a rough sketch I had made, wherein a little inaccuracy occurred which I now correct. It will be perceived that for want of room to place the spring (6, fig. 5), the screws which affix it to the Slide Plate enter into the grooved edge. This must not be, and I, therefore, suggest the substitution of a spring of a different shape, as hereagainst shewn. The

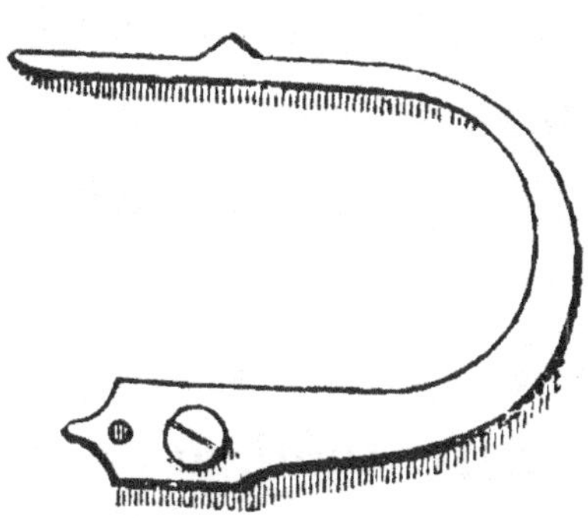

Clicks which take into and stop the centre wheel, are portions of small wheels, say of wheels of twenty-four teeth. Through the centre of these a screw passes, and attaches them to a plate of brass about an eighth of an inch thick, of any shape, such as is represented by the outline which encloses 4 and 5: This Click piece moves on a centre constructed as follows:

1. 1. is a piece of steel, with a hole drilled through it, to receive the screw 3. (the head is seen at 4. in plate fig. 5.) which attaches it to the Slide Plate.

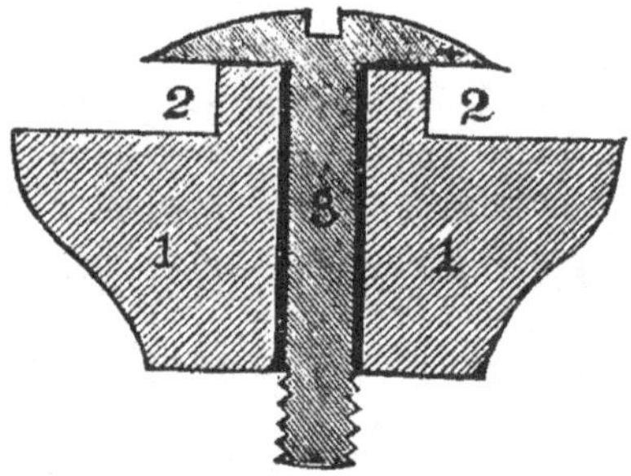

The Click piece is fitted nicely on the shoulder 2. 2. and the screw 3. confines it so as to let it move round on that shoulder without shake. This steel piece is of such thickness as will place the Clicks in the same plane with the centre toothed wheel, and allow space between the Click piece and Slide Plates, for the spring seen at 6, and roller seen at 7. The roller is hardened steel, and the spring, which should be rather a strong one, has a projecting ridge at its end, which acts against the roller. By pressing the end (5.) of the Click piece to either side, the spring will throw the Click, on the same side, into contact with the centre toothed wheel. Each Click has a screw through its centre, which attaches it to the Click piece; and one of them has, also, a screw which passes through an eccentric long opening in it, and clamps it fast to the Click piece in any required angular position. It is expedient that this Click, with the elongated eccentric opening, should be adjusted and clamped in its proper place in the progress of constructing the Chuck, before the circular movement is divided and figured; and I recommend the following method of proceeding. The Chuck being screwed to the mandril, the two Slide Plates must be adjusted so that they will traverse on their respective Bed Plates, in a parallel line with each other as near as may be, and the Click with the long opening is then to be placed in contact with the centre toothed wheel. The clamp screw of the Click must be loosened so as to allow the Click to move on its centre screw, it must also be ascertained that the Slide Plates *now* traverse *perfectly* in parallel lines to each other: and this point having been proved by two or three trials, the clamp screw in the long opening in the Click must be screwed quite tight, *not again to be moved* unless by any chance the Click should get out of this—its proper position. The place on the circle to which the index points, should be marked for the place of the *prime division,* which I shall further notice when I come to the description of the circular movements. I will here mention the method I employ to ascertain that the two Slide Plates traverse in perfect lineal parallelism. The Chuck being screwed on the Mandril, a piece of hard wood is fixed on it, and the surface of the wood is turned down to a plane. A pointed tool is fixed in the Slide Rest, the point touching the wood, and the screw of the Slide Plate, *next the Lathe head*, being turned, and the Slide Plate being thereby caused to traverse both ways on the Bed Plate, the point of the tool will describe a right line on the piece of wood. This Slide Plate is then to be brought to its central position, and the screw of the other Slide Plate is to cause the wood to move, backwards and forwards, in a right line against the point of the tool. If the point of the tool traces, in the second instance, the same line it delineated in the first, the Slide Plates, as a matter of course, must have traversed in perfect lineal parallelism. If *two* lines any where appear, the error must be got rid of by adjusting the circular movement, *next the Lathe head*, whilst the clamp screw of the Click is loose, as before mentioned.

The Click, at the other side of the Click piece, which has only a screw through its centre, is to be resorted to whenever an adjustment of the circular movement is required that cannot be obtained by the other Click alone. Suppose, for instance, a division of 192 be required, it cannot be obtained by one of the Clicks alone, but by the action of the two Clicks, conjointly, it can. The screw in the centre of the Click must be loosened, so that the Click will move on its centre, and the Click is to be brought into contact with the toothed wheel of the centre movement. The toothed centre wheel must now be moved, so as to cause the index to point exactly half way between any two of the divisions of 96, and then the screw in the centre of the Click is to be screwed fast. In this state of the Clicks, if they be brought, alternately, into contact with the toothed

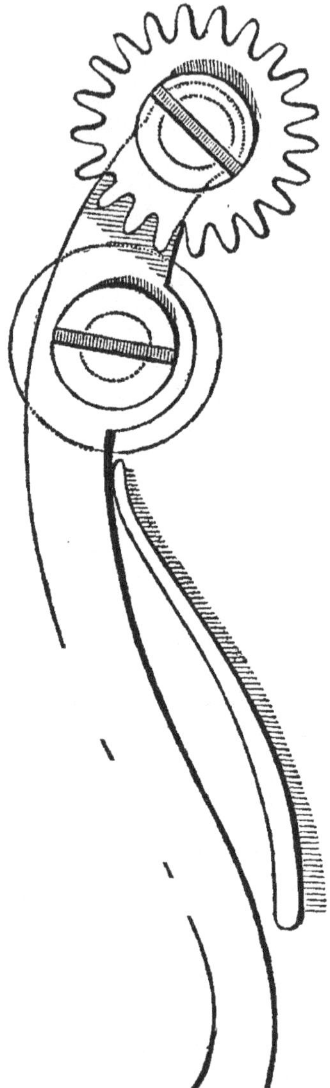

wheel, the centre movement will be moved only half a tooth at each adjustment, and thus a division of 192 will be obtained; and it is evident, that, by altering the angular position of this Click, the centre circular movement may be stopped at every possible point; and this Click is, therefore, most truly speaking, A MICROMETER ADJUSTING CLICK.

I have before mentioned that I had judged it expedient to remove the tangent screw and racked edge circle from my Chuck, and to substitute a click and toothed wheel. When, however, I decided upon doing this, I found that the common kind of click would not answer my purpose—that it was necessary to employ one that would adjust to any minute portion of *a tooth* of the wheel; and it was, therefore, on this occasion that I first contrived the click which I here give a plan of, and which I have denominated *The Single micrometer adjusting Click*; and this Click I now employ, in the Chuck

I work with, on the circular movement which is *between* the two Slide Plates, instead of the Double acting Click described at page 29. The adjustment of it is the same as that of the Click with the long opening (4, 5, fig. 5,) which has been already explained, and the necessity of using such a Click will be obvious when it is considered how very difficult it would be with the common Click to set the Slide Plates in perfect *lineal parallelism.* Without the power of parallel adjustment of the Slide Plates, and of retaining them in that position by means of the action of the Click on the toothed wheel, the Chuck would be very defective in its work: if they cannot be set parallel, they cannot be set at *right angles,* or at any other angle required; and thus squares, and triangles, and polygons, could not be correctly accomplished—indeed, there are so many figures and beautiful patterns that depend upon the correct angular adjustment of the Slide Plates, that a micrometer adjusting Click, or something that will answer the same purpose, is actually indispensable.

The division, on the divided circle of the circular movement, to which the index points when the Slide Plates are in lineal parallel position, and which division I have, in another place, called the *prime division*, is to have the number 12 engraved over it. I have now, in the course of long practice, adopted and used, on various occasions, the simple and plain notation of the divisions of a circle which is shewn in fig. 1. The toothed wheel consists of 96 teeth, and every *eighth* tooth is distinguished, consecutively, by the numerals 1, 2, 3, 4, 5, 6, 7, 8, 9, 10, 11, 12. In the Plate there is only a divisional line to every *two* teeth, but, with reference to the *Micrometer Click*, before spoken of, I recommend that a divisional line be placed for *every* tooth in the wheel, and that the space of one tooth, say, the one next to division

12, shall be divided in half—that the space next to that be divided into three parts—and the next space into four parts, and I give a cut hereunder to explain exactly what I mean; and this engraving I executed by means of my Compound Chuck.

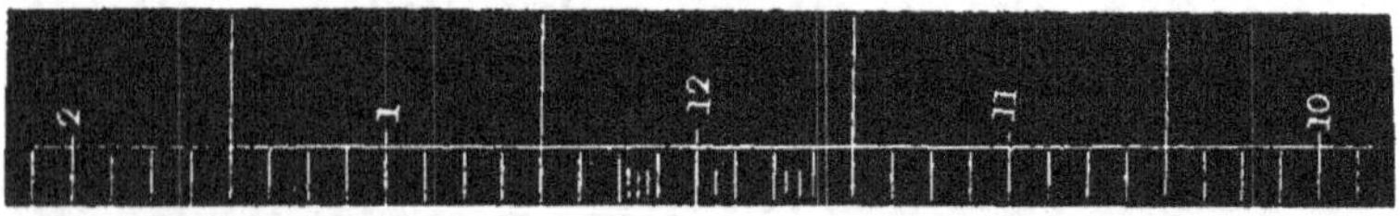

Practice has taught me that the fewer the figures, the less frequently will mistakes arise, and the less puzzling will be the adjustment of the circular movements.

The scale on the edge of the Bed Plates (see Plate I, fig. 1, at B. B.) should be so divided that *every turn* of the Slide Plate screw shall cause the index to point to a divisional line; the grand divisions are to be distinguished by a numeral, and each of them is to contain ten of the smaller divisions; and, when the circular movements (see K. K.) are in the middle of their respective Bed Plates and concentric with the Mandril, the index is to point to the divisional line on the scale which has 0 engraved over it.

It has been already suggested that the Slide Plates are to act in perfect lineal parallelism when the index, to that circular movement which is between them, points to the division marked 12, and from thence it follows, that they will also act in lineal parallelism, when the index points to division 6. When the index points to division 3 or 9, they will act at right angles to each other. When the index points to division 2, 4, 8, or 10, their action will be at an angle of 60; and so on for their acting at any other angle.

The Manuel du Tourneur was the first book that gave any account of an Eccentric Chuck; and I do not now know of any other book that does so. That book gives, also, a description of a *Compound* Eccentric Chuck; but according to the construction of it therein set forth, it is an instrument very limited in its action and powers. The Slide Plates have no circular movement between them, but are *fixed* to act at right angles to each other; and they, moreover, can only be thrown out of centre at *one* end of the Bed Plate; but the discrepances which arise from this construction of the Chuck will, however, more fully appear from the Specimens that will be herein given of the action and powers of a Compound Eccentric Chuck constructed on a different plan.

It appears from the title page of the edition of the Manuel du Tourneur which I possess, that it was published by Bergeron in 1792; and from that period the defects I have mentioned, in the construetion of the Compound Eccentric Chuck, have continued unheeded by the manufacturer of Turning Apparatus. A new Compound Eccentric Chuck, just finished by one of our first-rate London manufacturers, has been shewn to me, which, as heretofore, partakes of these and other most glaring defects.

Figure 3, is a Section—supposed to be through the centre of the Chuck, and coincident with the centre of the Mandril,—showing the internal construction of the circular movements; and, in describing this Section, I will detail the progress pursued by myself in the manufacture and putting together of the several parts of the Chuck.

The Boss of the Bed Plate, A. fig. 1, being screwed and fitted on the screw of the Mandril, a hole was turned out in the centre of it equal in diameter to the width of the long opening *c. c.* to be made in the Bed Plate. The face of the Bed Plate was then turned down, with the Slide Rest, and cleaned off with fine emery paper to a true plane; and the ends of the Bed Plate were also turned off with the Slide Rest. The Bed Plate was now unscrewed from the Mandril to be chucked again in the following manner, in order to the back part of it being turned and polished off. I have brass flanch Chucks, with holes through the flanch part, for screws to pass, to attach circular pieces of mahogany to them about an inch in thickness. One of these Chucks was screwed on the Mandril, and the face of the mahogany countersunk with the Slide Rest, to receive, *tightly*, the face and ends of the Bed Plate; and the Bed Plate was held fast in this countersink by placing a circular piece of plate brass (with a hole at the centre to pass a screw) on the Boss A. and screwing it to the wood. The Boss A. and back of the Bed Plate was now turned off and polished. The Bed Plate was then unchucked, and the holes drilled and screwed for the steady pins and screws which attach the steel arms to the Bed Plate, also the screw holes, in the edge of the Bed Plate, were drilled and screwed to receive the screws *i. i.*; and the long opening *c. c.* in the Bed Plate was worked out by drilling and filing. The steel arms with their angular edges, having been got up with the greatest care, so, as far as possible, to secure the Slide Plate to traverse between them in perfect parallelism with the surface of the Bed Plate, are to be screwed in their places to the Bed Plate. The Slide Plate having been prepared, and its faces made parallel, and its edges grooved and nicely fitted to slide between the steel arms, is to be placed between them on the Bed Plate, and, by the adjusting screws *i, i,* made to move rather tightly. Oil-stone dust and oil was now, from time to time, applied to the grooves, and the Slide Plate moved backwards and forwards on the Bed Plate until the angular edges of the steel arms and the grooves of the Slide Plate became a *perfect fit* by grinding. During this grinding process the screws *i. i.* were used, from time to time, to keep the Slide Plate in rather tight contact with the steel arms.

Let it now be supposed that every thing exhibited in figure 5 is finished and attached to the Slide Plate and Bed Plate by their respective screws; and that the next thing to be done is to drill the holes through the pieces D. and *f.* fig. 1, preparatory to fixing the Slide Plate screw *d. d.* in its place. It is, evidently, most essential to the good and correct working of the Chuck, that this screw should lie, when in its place on the Bed Plate, in a perfectly parallel line with the grooved edges of the Slide Plate and midway between them; and that to accomplish this the greatest care must be taken in drilling these holes. The Slide Plates of the Chuck I work with happen to act with great accuracy; and, therefore, under the idea of its being acceptable to the Amateur, I will detail the method I pursued in adjusting the Slide Plate screws.—I first ascertained the distance that the screw *d. d.* should range from the surface of the Bed Plate, (see figure 1) and I filed up a bit of brass plate to the exact thickness, that, when it was laid on the face of the Bed Plate, the edge of its upper surface extended exactly to where the centre of the screw *d. d.* should be. This bit of brass served then as a gauge, and by the aid of it I drew a fine line across the outside faces of the two pieces D. and *f.* which marked, exactly, the distance from the face of the Bed Plate where the holes were to be drilled. The next thing was to ascertain and make a mark in each of these lines exactly equidistant from the grooved edges of the Slide Plate, and midway between them; and I did not succeed in doing this to my mind, until I had contrived and made a tool, or gauge, expressly for the purpose. I give a description of this gauge, as it enabled me to accomplish a thing which I had previously despaired of doing.

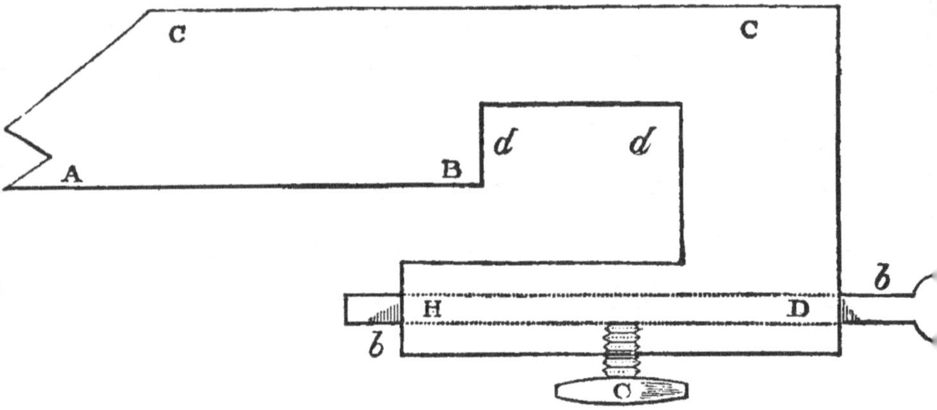

The piece **A. B. C. D.** is of brass plate, rather more than the eighth of an inch thick. The angle at **A.** fits the angular edges of the steel arms E. E. F. F. fig. 5, and when it is applied to one of these edges, the edge from **A.** to **B.** rests on the face of the Bed Plate **B.** figure 5, and the space *d. d.* passes over the other steel arm and brings the part **D. H.** to the opposite face of the Bed Plate. A pin of steel *b. b.* fits into and passes through a hole drilled through the part of the gauge as seen at D. H. and is held fast in its required position by the clamp screw C. The extreme end of the pin at H. *b.* intersects the line which had been previously drawn across the projecting pieces D. and *f.* on the Bed Plate fig. 1, and by repeated trials —first applying the angle A. to one steel arm and then to the other, and marking with a pencil the point at which the end of the pin crossed the line—I, at length, adjusted the end of the pin so that its extreme edge pointed exactly midway between the two steel arms, and then with a fine point I marked the place on the pieces D. and *f.* At the points where the cross lines intersected, which had been ascertained in the manner I have just explained, and marked on the pieces D. and *f,* dots were now made with a centre point, and holes were drilled through these pieces from those dots, in the following manner:—The drill stock was screwed on the Mandril, with a drill in it of the proper size for the screw hole to be drilled through D, and a pointed centre was placed in the cylinder of the front centre head. The dot in the piece D. at the intersection of the cross lines, was placed to the point of the drill, and the dot in the piece *f.* to the point of the front centre head; and the hole was drilled through D. by turning the Lathe round, and gradually screwing forward the cylinder with the pointed centre. The hole through the piece *f.* was then drilled with a proper size drill; the centre point in the front centre head being, in this case, pressed into the hole that had been drilled through the piece D.

By placing the screw *d. d.* in the situation and manner here represented, the Slide Plates may be thrown out to *either end* of the Bed Plate, thereby giving to the Chuck the power of forming Squares, and of executing a great deal of beautiful work, which no Chuck could do that is constructed on the ordinary plan of *throwing out at only one end.* These screws, moreover, should be of the same velocity as the screw of the Slide Rest; that is, they should have the same number of threads to an inch; and, in making them, every possible care must be taken to secure them against being, in the least, what is called *drunken*. I worked out these screws in the following manner, in the Lathe with the screw stock and dies, which I have found to be a very effectual way of making a good and correct screw. Having procured two pieces of round bar steel, sufficient in diameter to form the shoulder seen at *f*. above the letter B, I filed them up to the exact length required for the screw; and, having centre-pointed them at each end, a conical hole was drilled at one of their ends to receive the point of the screw, as seen at D. These pieces were then chucked between two centres, and turned down, in the Lathe, in the usual way, to the proper size to form the screw, finishing off the shoulder and end beyond it, to the dimensions as seen at *f*. The screw dies of the proper thread for the screw *d*. *d*. being placed in the screw stock, one of the pieces of steel prepared for the screw, as just mentioned, was put between the dies, and again chucked in the Lathe between two centres. The Lathe being turned round, the die stock and dies were caused to traverse and form a screw-thread from one end of the steel piece to the other; and by continually working the dies, backwards and forwards, in this way,—using plenty of oil,—and screwing up the dies by little and little, as occasion served, the screw was completed. These screws cannot, however, be finished off until after the gun-metal rings (hereafter described) are screw-tapped and finished, as they must be made to fit, or screw through those rings in the most perfect manner. Throughout the process of making these screws, the die stock must be kept firm and steady in the two hands, and be caused to traverse at *right angles* with the screw that is being made. If the die stock be allowed, in any part of its progress, to incline to one side or other of the screw-piece, the screw will, in that part, be imperfect—

will partake of the defect called drunkenness.

The Slide Plate is now to be adjusted on the Bed Plate, exactly in the middle, between its two ends; and, whilst in this situation, a hole is to be drilled through them both, at one corner, about half an inch from the end of the Slide Plate and the grooved edge. This hole is to receive a steady pin, to afford the means of adjusting and securing the Slide Plate in the same central position whenever required. The Bed Plate being now screwed on the Mandril, and the Slide Plate being secured in its central position by the steady pin, a hole is to be turned through the Slide Plate of the size required to receive the plug piece N. N. fig. 3, and a countersink is to be turned out round this hole to receive the shoulder of the piece P. P, as seen in the engraving fig. 3. The Slide Plate may now be removed from the Bed Plate, and the holes drilled in it, screwed, and finished off for the screws which belong to, and fix the index *t. t,* fig. 1, and the index 3, the click 4, and the spring 6, fig. 5.

The piece P. P, the plug piece N. N, and the screw-nut *k. k,* fig. 3, are made of steel; and the cylindrical piece or ring *q. q.* fig. 3 and 4, is made of gun-metal. The piece P. P. must be correctly turned out to fit, tightly, into the countersink in the Slide Plate. The cylindrical part of the plug piece N. N. next its screwed end, must fit *tightly* into the hole in the centre of the piece P. P, and must pass, *loosely,* through the hole in the centre of the Slide Plate; thus, by the action of the screw-nut *k. k,* the piece P. P, the Slide Plate G. G, and the plug piece N. N. will be attached as firmly together as though they were formed out of one piece of metal. The part of the plug piece N. N. seen at 4, must be finished off to a true cylinder, and the gun-metal ring *q. q.* must be made to fit it, rather tightly, in the most correct manner; and the ring being then removed from the plug, the holes are made in them to receive the screw *d. d.* To make these holes correctly, I had two drills, which run true in the Lathe, of sufficient length to pass through the pieces D. and *f.* fig. 1, and reach to the plug; the shank, between the stock and the point of one of these drills, exactly fitted the screw-hole in the piece D. and the shank of the other, the hole in the piece *f.* When the drill was passed through the screw-hole D. the front centre point was pressed into the hole in the piece *f.* and the Lathe turned round until the plug became drilled half through. The drill which fitted the hole in the piece *f.* was then placed in the stock, and the front centre being pressed into the screw-hole in the piece D. and the Lathe turned round, the plug became drilled through its other half, and as this was the largest of the two drills, it was passed quite through the plug. This hole in the plug was afterwards broached out, so as to leave plenty of room for the screw *d. d.* to pass through it without touching; and any burr that was found on the edges of the hole was carefully removed. The gun-metal ring was now placed on the plug, and the places marked on it where it was to be drilled and screw-tapped to receive the screw *d. d.* This was done, by means of the two drills, in the same way as the hole was drilled in the plug; but the drills were not worked into the ring more than about two-thirds of the ring's thickness; because, if the drill were passed quite through the ring, a burr would be thrown up on the inside next the plug, which would render it impossible to remove the ring from the plug. The ring, therefore, is to be taken off the plug (a mark being first made on the ring to record its proper position on the plug), and the holes are to be completed through its two sides, with a drill, of the proper size for the screw *d. d.* fixed in the drill stock screwed on the mandril, the ring being pressed against the drill by means of the front centre point, which is to be adjusted in the hole on the opposite side. The screw-taps, corresponding with the screw *d. d,* are to be passed through these holes in the ring, from side to side; and the screw-holes being completed with the finishing tap, the screw *d. d.* can be now finished off (as before mentioned), so that the ring may work freely, but tightly, from one end of it to the other. The screw *d. d.* having been worked through the screw of the ring from end to end, two or three times, with plenty of oil, the ring was taken off the screw, and any burr the screwing had thrown up inside the ring, was thoroughly filed away with a *riffler* file,* as the least burr would of course be an impediment to the ring being placed on the plug.

The Bed Plate being screwed on the

* Riffler Files are variously shaped. The one I used was bent, so that the burr could be removed without defacing, in the least, the inside surface of the ring.

Mandril, and the Slide Plate secured in its place by the steady pins, the piece P. P. is to be turned truly over and shaped as seen in fig. 3, and the holes drilled in it and screw-tapped to receive the screws *s. s.* The piece K. K. (made out of well hammered brass) is to be fitted most correctly, by turning and a little grinding with oil-stone dust and oil, to move round the piece P. P. The piece which the screws *s. s.* clamp to the piece P. P. is turned out of a ring of steel, and by causing the flanch rim of it to protrude beyond the top edge of P. P. and extend over part of the top surface of K. K, the latter is confined and kept firm to circulate, without the least shake, on and round the piece P. P. The outside surface of the brass ring piece K. K. is divided into 96 divisions, and figured as already explained. The piece L. L. is made out of well hammered brass, and the edge of it is cut into 96 teeth as seen in figure I, and it is shouldered into the piece K. K. in the manner seen in figure 3, to which it is screwed by four screws *o. o.* The part 3. 3. and other space not shaded with lines, holds oil, none of which can possibly escape, excepting between the bevelled surfaces of K. K. and P. P. being those which require it. For lubricating the surfaces that are in moving contact, such as the surface in question, I use neat's-foot oil, which I thicken by grinding with it a small portion of black-lead.

In figure 3. the toothed wheel piece L. L. has a projecting part, which constitutes the nose screw of the Chuck, as seen in figure 1, and the whole, from the surface of the Slide Plate to the top of the nose screw, I designate as "the Circular Movement of the Second Bed Plate."

In figure 1, the toothed wheel piece L. L. forms part of the "Circular Movement of the First Bed Plate;" and, as there seen, it is screwed to the projecting part at the back of the second Bed Plate. This projection at the back of the second

Bed Plate forms one casting with the Plate, and is turned off, under the same plan of chucking, &c. as has been suggested regarding the Boss A. at the back of the first Bed Plate. A shallow countersink is turned out in the toothed piece L. L. into which the projection of the second Bed Plate, just mentioned, must correctly fit, and the Bed Plate and the piece L. L. are then to be firmly screwed together by four screws passing through the Bed Plate at its projecting part, and screwing into the toothed piece L. L. the heads of these screws being, of course, countersunk in the Bed Plate.

After I had constructed my Chuck on the plan I have now detailed, and according to the representation of it in the engraving fig. 1, and had done a great deal of work with it, and had proved it to be a very efficient instrument, I yet thought that greater steadiness might be obtained by enlarging the diameter of the circular movement of the first Bed Plate. I, ac-

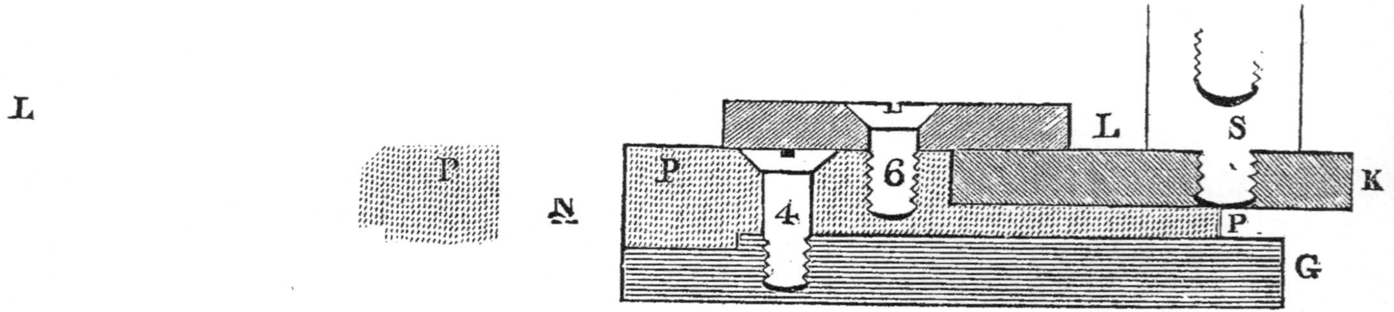

L
P
N
P
4
6
L
S
K
P
G

cordingly, made the alteration; and as my object is to relate all I know on the subject of the Chuck, I give a diagram, of this alteration, in the real dimensions of the pieces which compose it.

The annexed wood-cut is a section through the centre, shewing a modification of the construction of the circular movement and its appendages, which I have recently introduced into my Chuck on the Bed Plate next the Lathe head, in substitution of the one represented by the engraving, figure 3. By doubling the diameter of the moving circular plate, additional steadiness is obtained; and the power of angular adjustment is greatly increased by having a stop wheel of 192 teeth instead of one of 96; and, besides, the construetion is more simple and easy. G. G. is the Slide Plate, made of brass, attached to the Bed Plate that screws on the Mandril. A piece of steel P. P. P. P. is shouldered into this Slide Plate and kept firm to it by four screws (see two of them marked 4). N. represents the circular hole passing through the Slide Plate and steel piece, to receive the plug N, described before and shewn in the engraving figure 3. A circular brass plate K. K. the edge cut into 192 teeth, is fitted to move round, without the least shake, on the steel piece P. P. P. P. A steel plate L. L. is screwed by six screws (see those marked 6) to the steel piece P. and the brass circle with teeth is kept firmly in its place by it. The centre part of the plate L. L. is turned out to make room for the plug screw nut, figure 3. N. Into the brass plate K. K. there are four pillars S. S. screw quite tight, at an equal distance from the centre of the plate and from each other. O. O. is a plate of brass screwed firmly to these pillars, in the manner seen. When this plate is screwed to the pillars the edge is turned truly round, and divided and figured into twelve grand divisions, and each of those divisions, again, into eight parts. A slender piece of brass plate is screwed to the back of the steel piece P, on one of the sides that project beyond the Slide Plate, to carry the index to the circular division. A piece of brass plate is also attached, by screws, to the back of the piece P. to carry the micrometer click (before described) that is to act on the toothed edge of the plate K. The space which is obtained by means of the pillars, between L. L. and O. O. is for the Slide Plate screw, which, in the engraving fig. 1, is seen next to the toothed wheel L. L. The second, or upper Bed Plate, shewn in fig. 1, and represented in fig. 5 by B. B. is, under this modification of the Chuck, substituted by O. O. The steel arms E. E. F. F. seen in figure 5, are attached to O. O. in like manner, as they therein appear; but, as O. O. is a circular plate, the adjusting screws *i. i.* cannot be used; and, therefore, two snail-headed screws may be substituted, or a bar of brass, of the thickness of the adjustible steel arm, may be screwed to the plate parallel to the arm, and two screws passed through it, in a plane with the plate, which when screwed will press against the steel arm as occasion may require. In my Chuck I have adopted the latter.

Having now gone into an explanation of every thing that particularly occurs to me, with regard to the Compound Eccentric Chuck, the next thing will be to give the Specimens of simple Eccentric Turning which have appeared in the first and second editions.

INTRODUCTION

TO THE FORMER EDITIONS.

TURNING is now so generally resorted to as an amusement, that the communication of any thing, which may tend to produce novelty in the art, cannot fail of being interesting to the Amateur; besides, there can be no better means employed to promote improvement than circulating the result of experience.

Turning has been the favourite amusement of the Author for many years, and from the time he first took to it, he has sensibly felt the advantage he would have derived, could he have informed himself of the progress which had been made by others, and he regrets the information that has daily escaped him, for want of a reciprocity of communication among his cotemporary Turners. With regard to himself, he has long had it in contemplation to make known the result of some of his labours; but the very great expense of getting the necessary engravings executed—indeed the almost impossibility of getting them done at all—has hitherto prevented him from carrying his intention into effect. He has, however, in pursuing his object, constructed, himself, a piece of Machinery which enables him to engrave his own copper-plates.

Having so far overcome his difficulties, he ventures to offer to the attention of the

Turner, a Series of Specimens in Eccentric Circular Turning; and should this be the means of inducing others to communicate the result of their labours to the Public, in the same or any other branch of Turning, he will be much gratified.

The copper-plate Engravings represent the pieces of Eccentric Circular Turning to be executed on ivory, wood, or any other material; and each plate is accompanied by such Practical Instructions, as will enable any one, who has acquired the least practical knowledge in the art to produce the desired effect.

Plate 1.

SPECIMENS

IN

Eccentric Circular Turning.

SPECIMEN I.

THE pattern in Eccentric Turning which is described in this Specimen, and represented in the accompanying plate, is peculiarly adapted for the lid of a snuff-box; or three or four of the centre figures may be selected, and applied to decorate a set of backgammon-men.

This Specimen is composed of seven sets of circles, of different radii, arranged, at different eccentricities, round the common centre. The number of circles which compose the interior, or centre set, is twelve; the number in the exterior set, is two hundred and eighty-eight: the latter number may, however, be greater than can be obtained with the common eccentric chuck, and another set, therefore, is described, which may be substituted by those who have not the means of adopting the one exhibited in the plate.

The following is a practical description of the method to be pursued to produce a piece of Eccentric Turning similar to Plate I.

The tools with which the work is to be executed are double angular, Nos. 25, 32, and 36, according to Messrs. HOLTZAPFFEL and DEYERLEIN's construction of this sort of tool;[1] and the slide rest is to be

[1] It has been often suggested to me, that it would have been better to have given a particular description of these tools. My answer to which has always been, that, it formed no part of my plan, when I published these Specimens, to give any account of the construction of the machinery. I considered that the amateur who was disposed to engage in the execution of any of these Specimens, must

set at right angles with the mandril of the lathe. Fix an angular tool, No. 25, in the slide rest, and adjust the eccentric chuck, slide rest, and the point of the angular tool, with the nicest correctness, to the common centre, and proceed to execute the interior, or centre set of circles.

necessarily be acquainted with the tools, usually employed in this branch of Turning, and that it was, therefore, sufficient to refer him to the manufactory where I knew they could be obtained, and where I always procured them myself.

FIRST SET OF CIRCLES.

[2] Eccentricity equal to 3.
Radius equal to 2.

To produce the eccentricity, turn the slide screw of the eccentric chuck backwards three turns; to produce the radius, turn the slide rest screw backwards two

[2] The *Eccentricity and Radius* are noted, in this manner, throughout. My object in doing this has not, I find, been exactly understood. It is this: All such figures depend solely on the ratio which the *Eccentricity* bears to the *Radius*. In this case the ratio is as 3 to 2; and this being known, the same figure may be produced of any size; and the same rule applies to all other cases. By attending to this the patterns may be varied, in size, at pleasure; and thereby the different figures, forming these specimens, may be combined in various ways, and infinite variety of new patterns may be produced. For instance: If it were required to place this figure in lieu of the centre figure, in Plate II.—As it is, it would be too large; but if the eccentricity be produced by $1\frac{1}{2}$ turns of the screw, and the radius by 1

turns;[3] then describe twelve circles, equidistant from each other, round the common centre. These circles must be cut sufficiently deep to give a sharp edge to the figures, which the angular tool will produce; and this effect must be obtained, although it may be attended with the labour of going over the circles, and cutting them deeper, a second or third time: but to whatever depth this set of circles be cut, the same depth must be preserved throughout all the other parts of this pattern, excepting the exterior set of circles, regarding the depth of which, mention will be hereafter made.

turn, the figure will be the same, though of only half the size, and would then be small enough to be introduced as the centre figure in Plate II. It might also, on the same principle, be enlarged. The eccentricity may be produced by turning the screw $3\frac{3}{4}$ turns; and the radius, by turning the screw $2\frac{1}{2}$ turns. The ratio, in this case, is still as 3 to 2; but the figure becomes increased in size.

[3] It is to be understood that the slide screw of the eccentric chuck, and the screw of the slide rest, are equal in velocity.

SECOND SET OF CIRCLES.

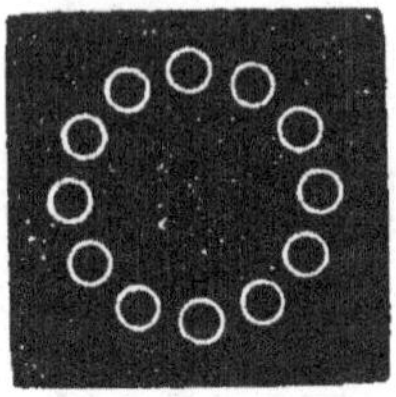

Eccentricity equal to 6.
Radius equal to 1.

To produce the eccentricity, turn the slide screw of the eccentric chuck backwards three turns; which, in addition to the former, makes the eccentricity equal to six.[4] To make the radius equal to one,

[4] I have adopted the expressions of "turning the screw backwards," and "turning the screw forwards," as being the matter of fact, and the precise things that must be done to produce the eccentricity and radius. And as every one must know what is meant by turning a screw backwards or forwards, I adopted those expressions, as being *more intelligible* than any other I could make use of.

I am induced to offer this explanation, in cousequence of the following paragraph, which appeared in a little amateur production, published two years after mine. The paragraph alluded to runs thus: The expression "*advancing the point*," though not turn the slide rest screw forwards one turn, which is a decrease of the former radius equal to one turn. Then with the angular tool as before, describe twelve equidistant circles.

THIRD SET OF CIRCLES.

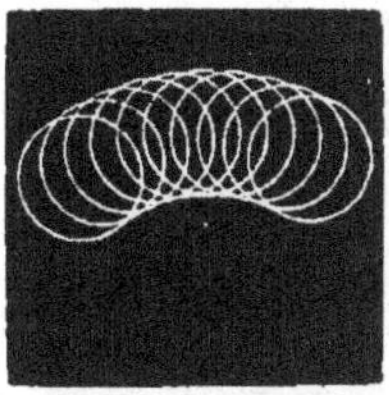

Eccentricity equal to 10.
Radius equal to 3.

Change the angular tool to No. 36, but, before it is fixed in the slide rest, adjust it to the radius and depth of the circle which was last turned. Then, to produce the eccentricity, turn the slide screw of

strictly correct, has been adopted in preference to "*moving the small wheel backwards,*" as more intelligible.

the eccentric chuck backwards four turns; and, to produce the radius, turn the slide rest screw backwards two turns. Describe forty-eight equidistant circles.

FOURTH SET OF CIRCLES.

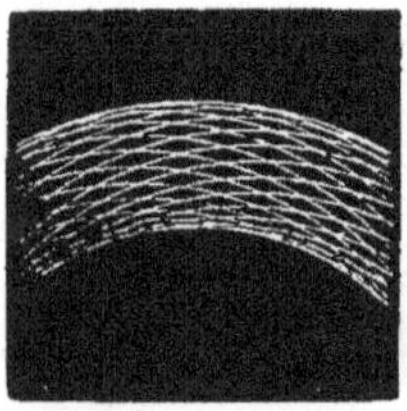

Eccentricity equal to 3.
Radius equal to 16.

Proceed with the same angular tool, and, to produce the eccentricity for this set, turn the slide screw of the eccentric chuck forwards seven turns. To produce the radius, encrease the former radius by turning the slide rest screw backwards thirteen turns. Then describe twenty-four equidistant circles.

FIFTH SET OF CIRCLES.

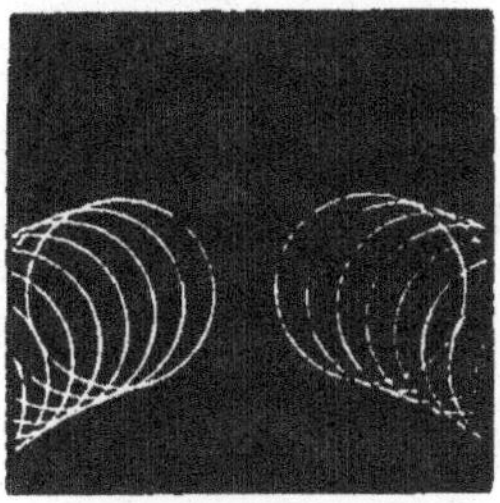

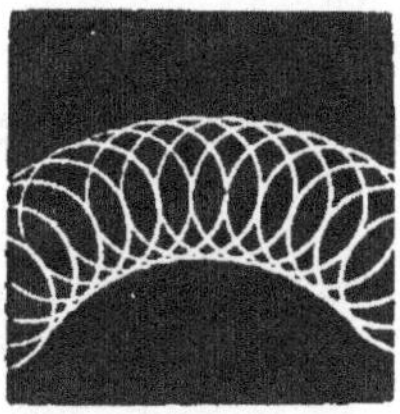

Eccentricity equal to 23.
Radius equal to 4½.

Continue with the same angular tool. The eccentricity is to be encreased by turning the slide screw of the eccentric chuck backwards twenty turns; and the radius is to be decreased by turning the slide rest screw forwards eleven turns and a half. Then (supposing the eccentric circle to be divided into ninety-six,) begin at eccentric division No. 96, and describe sixteen circles: miss eight divisions, set the eccentric circle to division No. 24, and describe sixteen more circles: miss eight divisions, set the eccentric circle to No. 48, and describe sixteen more circles: miss eight divisions, set the eccentric circle to No. 72, and describe sixteen more circles.

SIXTH SET OF CIRCLES.

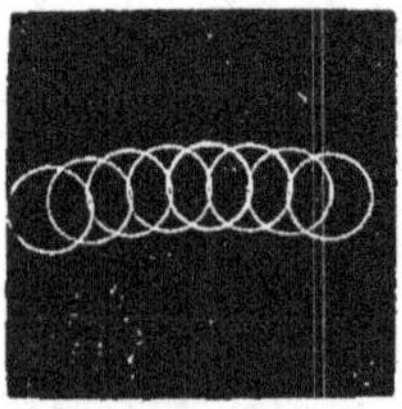

Eccentricity equal to 29½.
Radius equal to 2.

Change the angular tool to No. 25; which, before it is fixed in the slide rest, must be adjusted to the radius and depth of the circle last turned. The eccentricity is then to be encreased by turning the slide screw of the eccentric chuck backwards six turns and a half, and the radius is to be decreased by turning the slide rest screw forwards two turns and a half. Describe ninety-six equidistant circles.

SEVENTH SET OF CIRCLES.

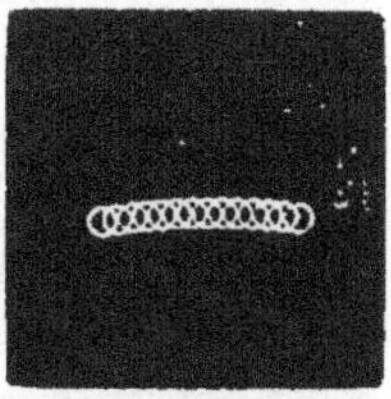

Eccentricity equal to 33.
Radius equal to $\frac{5}{8}$.

Change the angular tool to No. 32; and, before it is fixed in the slide rest, adjust it to the radius of the circles last turned. The eccentricity is then to be increased by turning the slide screw of the eccentric chuck backwards three turns and a half; and the radius is to be decreased by turning the slide rest screw forwards one and three-eighths of a turn. Describe two hundred and eighty-eight equidistant circles, not, however, of the depth of the circles previously turned, but of such a depth only as will just make the figures which the tool leaves, sharp at top. Or, for common eccentric chucks, the following may be adopted.

SEVENTH SET OF CIRCLES.

ADAPTED TO AN ECCENTRIC DIVISION OF NINETY-SIX.

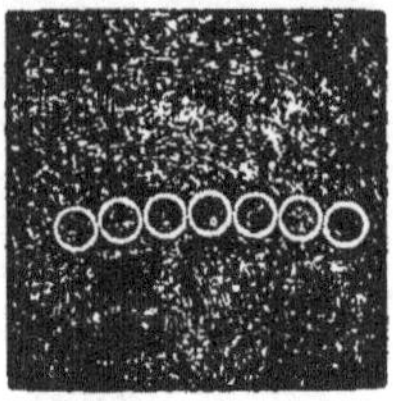

Eccentricity equal to 33.
Radius equal to 1.

Proceed exactly in the same manner as described for the set of two hundred and eighty-eight circles: the only difference is, that, in this set, the radius is to be made equal to one turn of the slide rest screw, and the number of equidistant circles must be limited to ninety-six.

Plate 2

SPECIMEN II.

THIS Specimen consists of seventeen sets of circles, of different eccentricities and radii, arranged round the common centre, in the manner represented in Plate II.

The exterior sets are produced by an eccentric circular division of two hundred and eighty-eight, and are formed into the figure of an Etruscan Border; but, as this high number of divisions cannot be obtained with the common eccentric chuck, a method is hereafter described, by which a Border, similar in figure, may be produced with an eccentric circular division of only ninety-six.

To produce a piece of Eccentric Circular Turning, similar to the pattern represented in Plate II, the following arrangement must be minutely attended to.

The work, in the first instance, is to be brought to a plane and polished surface; and the eccentric circles are to be executed, with double angular tools, Nos. 28 and 36. Set the slide rest at right angles with the mandril of the lathe, and let it remain in that position until the work be completed. Fix an angular tool, No. 28, in the slide rest, and adjust its point, the eccentric chuck, and the slide rest, to the common centre, with the greatest possible correctness; then proceed to execute the interior, or centre set of circles: these circles are not to be cut deep into the surface of the work; but must be cut sufficiently so to leave the figures sharp at top.

FIRST SET OF CIRCLES.

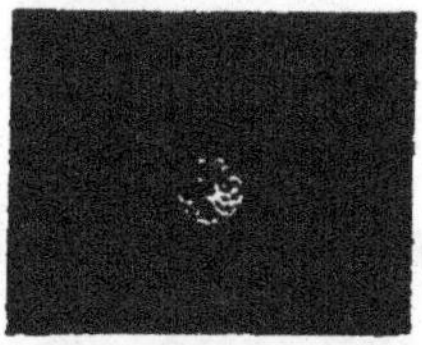

Eccentricity equal to 1.
Radius equal to $\frac{1}{2}$.

To produce the eccentricity, turn the slide screw of the eccentric chuck backwards one turn. To produce the radius, turn the slide rest screw backwards half a turn. Describe twelve equidistant circles.

SECOND SET OF CIRCLES.

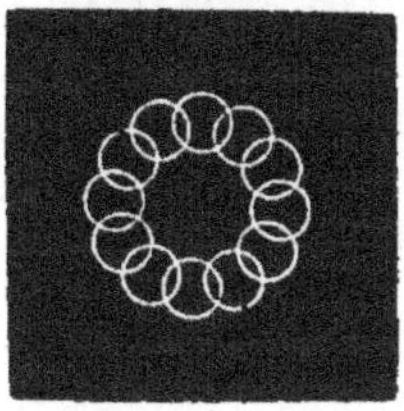

Eccentricity equal to 4.
Radius equal to $1\frac{1}{2}$.

To produce this eccentricity, increase the former eccentricity by turning the slide screw of the eccentric chuck backwards three turns; and to produce this radius, increase the former radius by turning the slide rest screw backwards one turn; then, with the same tool as before, describe twelve equidistant circles. These circles must be cut considerably deeper than the first set of circles; rather more so than will be sufficient to leave the figure, which is formed by the intersection of the circles, sharp at top: but to whatever depth these circles are cut, the same depth must be preserved in all the sets of circles remaining to be executed; excepting those composing the border, regarding which mention will be hereafter made.

THIRD SET OF CIRCLES.

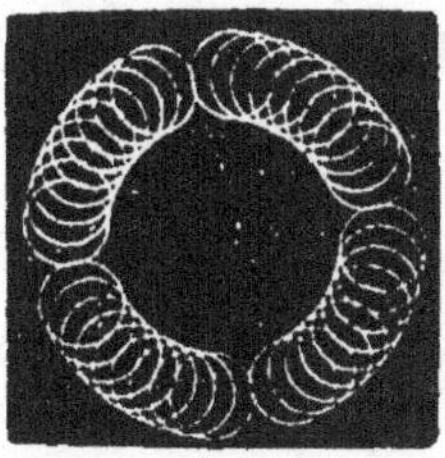

Eccentricity equal to $7\frac{1}{2}$.
Radius equal to 2.

Change the angular tool, and fix No. 36 in the slide rest; but, before it is fixed, adjust its point to the radius and depth of the circle last turned. Then produce the eccentricity by turning the slide screw of the eccentric chuck backwarks three turns and a half; and produce the radius, by turning the slide rest screw backwards half a turn. Describe thirty-two circles at the following distances: *viz.* (supposing the circle of the eccentric chuck to be divided into ninety-six,) begin at division 96, and describe eight circles, one circle at every alternate succeeding division: miss eight succeeding divisions, set the eccentric circle at division 24, and describe eight more circles, one circle at each alternate division: miss eight succeeding divisions, set the eccentric circle at division 48, and describe eight more circles, one circle at each alternate division: miss eight succeeding divisions, set the circle of the eccentric chuck at division 72, and describe eight circles as before.

FOURTH SET OF CIRCLES.

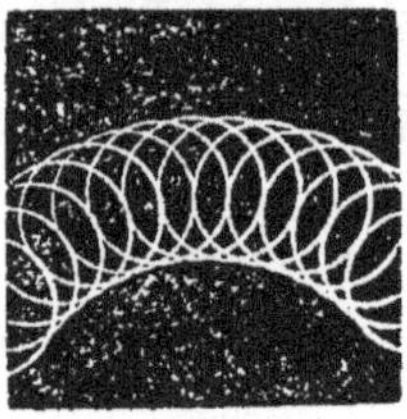

Eccentricity equal to 13.
Radius equal to $3\frac{1}{2}$.

Change the angular tool, and fix No. 28 in the slide rest; but, before it is fixed, adjust its point to the radius and depth of the circle last turned. Produce the eccentricity by turning the slide screw of the eccentric chuck backwards five turns and a half. Produce the radius by turning the slide rest screw backwards one turn and a half. Then describe forty-eight equidistant circles.

FIFTH SET OF CIRCLES.

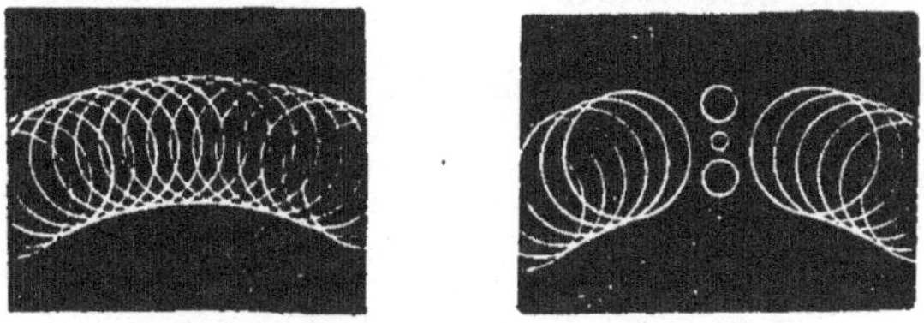

Eccentricity equal to $19\frac{3}{4}$.
Radius equal to $3\frac{1}{2}$.

Change the angular tool, and fix No. 36 in the slide rest; but, before it is fixed, adjust its point to the radius and depth of the circle last turned. The eccentricity is to be produced by turning the slide screw of the eccentric chuck backwards six turns and three-fourths: the radius remains as before. Describe sixty-four circles, and arrange them in the following manner: *viz.* Set the eccentric circle at division 96, and describe sixteen circles, one circle at each following division: miss eight divisions, set the eccentric circle at division 24, and describe sixteen more circles: miss eight divisions, set the eccentric circle at division 48, and describe sixteen more circles: miss eight divisions, set the eccentric circle at division 72, and describe sixteen circles as before.

SIXTH SET OF CIRCLES.

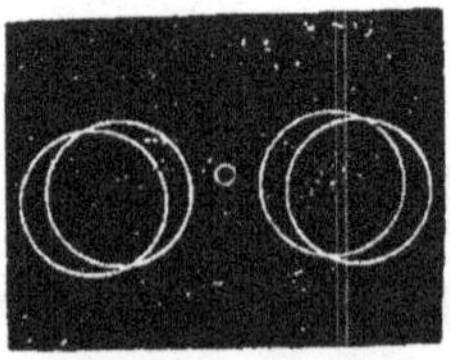

Eccentricity equal to $19\frac{3}{4}$
Radius equal to $\frac{1}{2}$

Proceed with the same angular tool, No. 36, and with the same eccentricity.

Decrease the radius by turning the slide rest screw forwards three turns. Then (supposing the eccentric circle to be divided into ninety-six) describe four circles, one circle at each of the divisions undernamed: *viz.* at division 20, 44, 68, 92.

SEVENTH SET OF CIRCLES.

Eccentricity equal to $21\frac{3}{4}$.
Radius equal to 1.

Proceed with the same angular tool, No. 36. Increase the eccentricity by turning the slide screw of the eccentric chuck backwards two turns: and increase the radius by turning the slide rest screw backwards half a turn. Then describe four circles, one circle at each of the following divisions, *viz.* at 20, 44, 68, 92.

EIGHTH SET OF CIRCLES.

Eccentricity equal to 17¾.
Radius equal to 1.

Proceed with the same angular tool, No. 36. Decrease the eccentricity by turning the eccentric slide screw forwards four turns; and, with the same radius as before, describe four circles, one at each of the following divisions, *viz.* 20, 44, 68, 92.

NINTH SET OF CIRCLES.

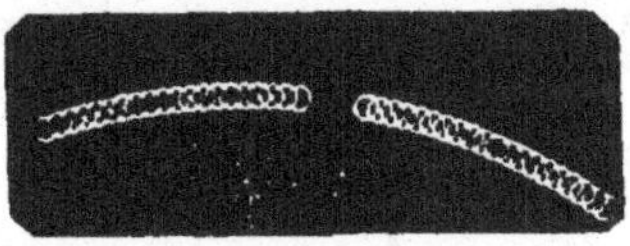

Eccentricity equal to 30.
Radius equal to $\frac{1}{2}$.

Change the angular tool, and fix No. 28 in the slide rest; but, before it is fixed, adjust its point to the radius of the circle last turned. The circles in this and the succeeding sets, which compose the Etruscan Border, are to be cut no deeper than will make the figures perfect at top.

Produce the eccentricity by turning the slide screw of the eccentric chuck backwards twelve turns and a quarter; and reduce the radius by turning the slide rest screw forwards half a turn.

The set of circles now to be produced, is the exterior of the nine sets which form the Etruscan Border. The circles in this set are two hundred and forty, which are to be described in the following manner: Set the circle of the eccentric chuck (which is supposed to be divided into two hundred and eighty-eight) at division 5, and describe twenty circles, one circle at that and at each succeeding division: miss four divisions, set the circle of the eccentric chuck at division 29, and describe twenty circles as before: miss four divisions, set the circle of the eccentric chuck at division 53, and describe twenty more circles; and so on for the rest of this set, which, if correctly executed, will be completed by describing the last circle, in the last series of twenty, at division 288.

TENTH SET OF CIRCLES.[6]

Eccentricity equal to $29\frac{3}{8}$.
Radius equal to $\frac{1}{2}$.

Proceed with the same angular tool and radius as before; and decrease the eccentricity by turning the eccentric slide screw forwards five-eighths of a turn. The circles in this set are twenty-four, and are to be described at the following distances. Set the circle of the eccentric chuck at division 5, and describe one circle: miss eighteen divisions, set the circle of the eccentric chuck at division 24, and describe one circle: miss four divisions, and describe one circle at the succeeding division; and so on, missing eighteen and four divisions alternately, and describing one circle at each division between those

[6] This and the following Illustrations exhibit the progress of the Etruscan Border; the lowermost circles being those which are to be produced.

two series, till the last circle being described at division 288, completes the set.

ELEVENTH SET OF CIRCLES.

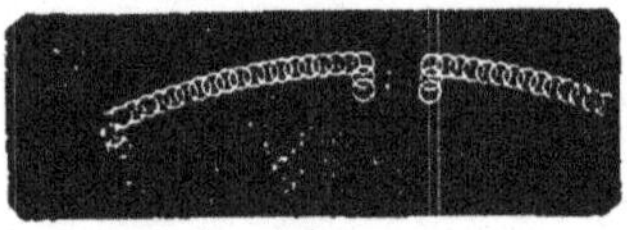

Eccentricity equal to 28$\frac{3}{4}$.
Radius equal to $\frac{1}{2}$.

Decrease the eccentricity by turning the slide screw of the eccentric chuck forwards five-eighths of a turn; and then proceed, with the same angular tool and radius as before, to describe twenty-four circles in the same manner as directed for the tenth set.

TWELFTH SET OF CIRCLES.

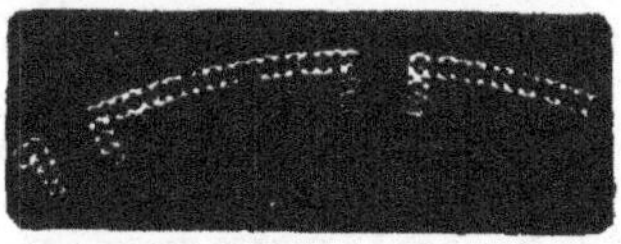

Eccentricity equal to $28\frac{1}{8}$
Radius equal to $\frac{1}{2}$

Decrease the eccentricity by turning the slide screw of the eccentric chuck forwards five-eighths of a turn; and proceed to describe twenty-four circles in the same manner as directed for the tenth set.

THIRTEENTH SET OF CIRCLES.

Eccentricity equal to $27\frac{1}{2}$.
Radius equal to $\frac{1}{2}$.

Proceed with the same angular tool and radius as before; and decrease the eccentricity by turning the slide screw of the eccentric chuck forwards five-eighths of a turn. This set consists of one hundred and eighty circles, which are to be described in the following manner: Set the circle of the eccentric chuck at division 5, and describe fifteen circles, one circle at that and at each succeeding division: miss four divisions, set the circle of the eccentric chuck at division 24, and describe one circle: miss four divisions, set the circle of the eccentric chuck at division 29, and describe fifteen circles: and so on for the remaining divisions of this set.

FOURTEENTH SET OF CIRCLES.

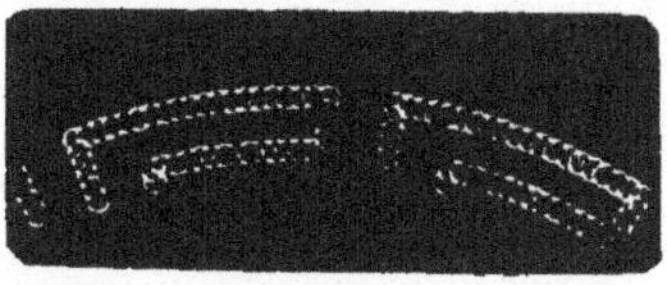

Eccentricity equal to $26\frac{7}{8}$.
Radius equal to $\frac{1}{2}$.

Proceed with the same angular tool and radius as before; and decrease the eccentricity by turning the slide screw of the eccentric chuck forwards five-eighths of a turn.

This set consists of twenty-four circles, which are to be described at the following distances: Set the circle of the eccentric chuck at division 288, and describe one circle: miss eighteen divisions, set the circle of the eccentric chuck at division 19, and describe one circle: miss four divisions, and describe one circle at the succeeding division: and so on, missing eighteen and four divisions alternately, and describing one circle at each division between those two series, till the last circle being described at division 283, completes the set.

FIFTEENTH SET OF CIRCLES.

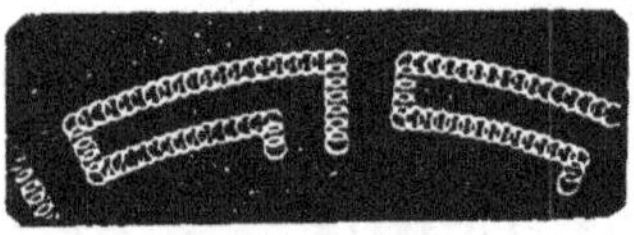

Eccentricity equal to $26\frac{1}{4}$.
Radius equal to $\frac{1}{2}$.

Decrease the eccentricity by turning the slide screw of the eccentric chuck forwards five-eighths of a turn: and then proceed, with the same angular tool and radius as before, to describe twenty-four circles, in the same manner as directed for the fourteenth set.

SIXTEENTH SET OF CIRCLES.

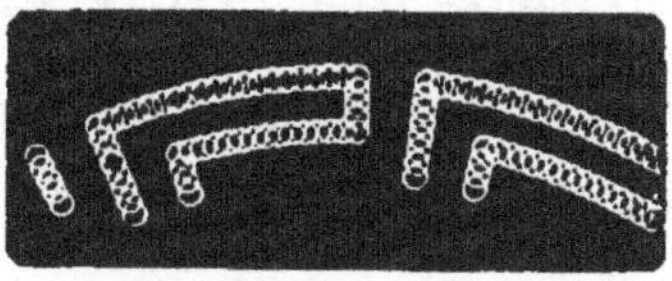

Eccentricity equal to $25\frac{5}{8}$.
Radius equal to $\frac{1}{2}$.

Decrease the eccentricity by turning the slide screw of the eccentric chuck forwards five-eighths of a turn; and then, with the same radius and angular tool as before, describe twenty-four circles at similar distances to those of the fourteenth set.

SEVENTEENTH SET CIRCLES.

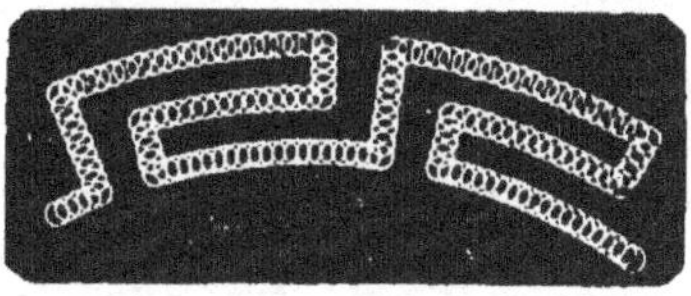

Eccentricity equal to 25.
Radius equal to $\frac{1}{2}$.

Decrease the eccentricity by turning the slide screw of the eccentric chuck forwards five-eighths of a turn, and proceed with the same angular tool and radius as before. This set of circles completes the Etruscan Border, being the interior of the nine sets of which that Border is composed. There is the same number of circles in this as in the exterior or ninth set, *viz.* two hundred and forty, which are to be described in the following manner: Set the circle of the eccentric chuck at division 288, and describe twenty circles, one circle at that and at each succeeding division: miss four divisions, set the circle of the eccentric chuck at division 24, and describe twenty circles as before: miss four divisions, set the circle of the cecentric chuck at division 48, and describe twenty more circles: and so on for the rest of this set, which will be completed, if correctly executed, by the last circle, in the last series of twenty, being described at 283.

The labour attending the execution of borders of this sort is very considerable, particularly when they are composed of so minute a circular division as that just described. This alone will probably be sufficient to induce attention to correctness in placing the circles at their assigned distances; for it must be recollected that the omission of a circle would considerably injure the general effect, and a circle misplaced would totally spoil it.[6]

By the following method an Etruscan Border, similar in figure to that which is represented in Plate II, may be produced with an eccentric circular division of only ninety-six. This Border consists of five sets of circles, which are to be cut sufficiently deep to make the figure perfect at top.

[6] At the time I wrote this I was unconscious I had afforded an opportunity of applying these remarks. In the copper-plate Engraving (No. 2.) I had omitted every alternate circle in the second set of circles in the Etruscan Border. This circle was, however, properly inserted in the wood-cut illustration (see 17th set of circles). The omission of it in the copper-plate is not only very apparent,

NINTH SET OF CIRCLES.

ADAPTED TO AN ECCENTRIC CIRCULAR DIVISION OF NINETY-SIX.

Eccentricity equal to 33.
Radius equal to $\frac{7}{8}$.

Change the angular tool, and fix No. 28 in the slide rest; but, before it is fixed, adjust its point to the radius of the circle last turned. Produce the eccentricity and radius, and make them greater than required for the ninth set, adapted to an eccentric division of two hundred and eighty-eight, by three turns backwards of the slide screw of the eccentric chuck, and three-eighths of a turn backwards of the slide rest screw.

but may, certainly, be said to "injure the general effect." It was not, however, discovered until after it was before the Public, and it will now serve as a beacon to those who may be desirous of perfecting such kind of work.

This set of circles is the exterior of the five sets which are, now, to form the Etruscan Border: the number of circles in this set is eighty-four, which are to be described in the following manner.

Set the circle of the eccentric chuck (which is supposed to be divided into ninety-six) at division 2, and describe seven circles, one circle at that and at each succeeding division: miss one division, set the circle of the eccentric chuck at division 11, and describe seven circles as before: miss one division, set the circle of the eccentric chuck at division 19, and describe seven more circles: and so on for the rest of this set, which, if correctly executed, will be completed by describing the last circle, in the last series of seven, at division 96.

TENTH SET OF CIRCLES.

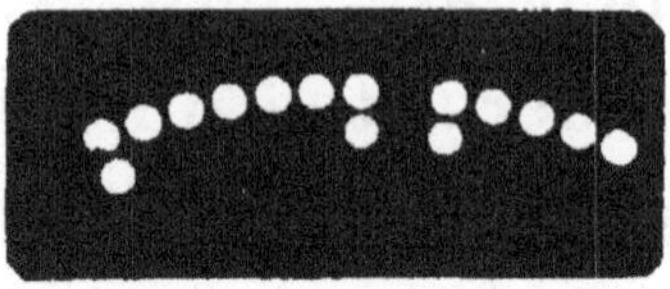

Eccentricity equal to $31\frac{1}{16}$.
Radius equal to $\frac{13}{16}$.

Proceed with the same angular tool as before. Decrease the eccentricity by turning the slide screw of the eccentric chuck forwards one turn and fifteen-sixteenths of a turn. Decrease the radius by turning the slide rest screw forwards one-sixteenth of a turn. The circles in this set are twenty-four, which are to be described at the following distances: Set the circle of the eccentric chuck at division 2, and describe one circle: miss five divisions, set the circle of the eccentric chuck at division 8, and describe one circle: miss one division, set the circle of the eccentric chuck at division 10, and describe one circle: and so on, missing five divisions and one division alternately, and describing one circle at each intermediate division, till the last circle, being described at division 96, completes the set.

ELEVENTH SET OF CIRCLES.

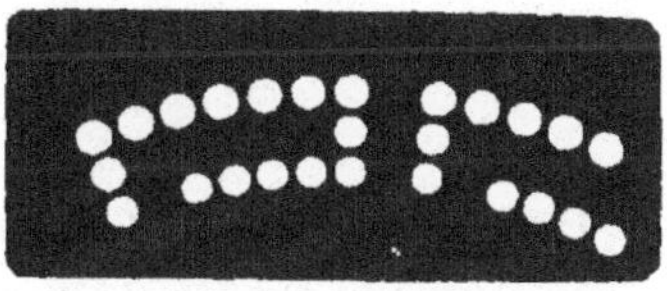

Eccentricity equal to $29\frac{3}{16}$.
Radius equal to $\frac{3}{4}$.

Proceed with the same angular tool as before. Decrease the eccentricity by turning the slide screw of the eccentric chuck forwards one turn and fourteen-sixteenths of a turn. Decrease the radius by turning the slide rest screw forwards one-sixteenth of a turn. There are seventy-two circles in this set, which are to be described in the following manner: Set the circle of the eccentric chuck at division 2, and describe five circles, one circle at that and at each succeeding division: miss one division, set the circle of the eccentric chuck at division 8, and describe one circle: miss one division, set the circle of the eccentric chuck at division 10, and describe five circles: and so on, till this set be completed.

TWELFTH SET OF CIRCLES.

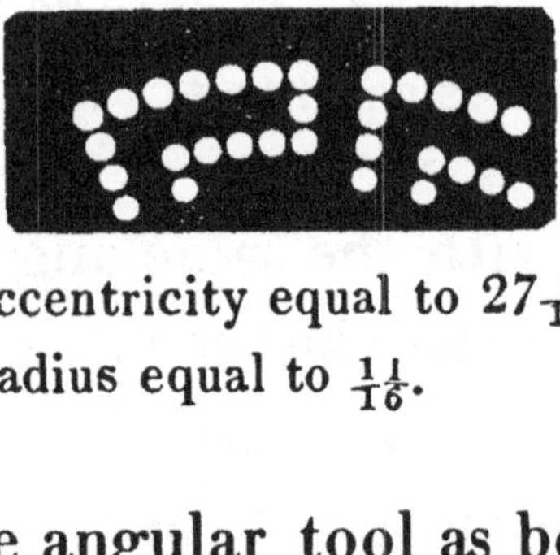

Eccentricity equal to $27\frac{6}{16}$.
Radius equal to $\frac{11}{16}$.

The same angular tool as before. Decrease the eccentricity by turning the slide screw of the eccentric chuck forwards one turn and thirteen-sixteenths of a turn. Decrease the radius by turning the slide rest screw forwards one-sixteenth of a turn. This set consists of twenty-four circles, which are to be described at the following distances: Set the circle of the eccentric chuck at division 96, and describe one circle: miss five divisions, set the circle of the eccentric chuck at division 6, and describe one circle: miss one division, and describe one circle at the succeeding division: and so on, missing five divisions and one division alternately, and describing one circle at the intermediate divisions, till the set be completed.

THIRTEENTH SET OF CIRCLES.

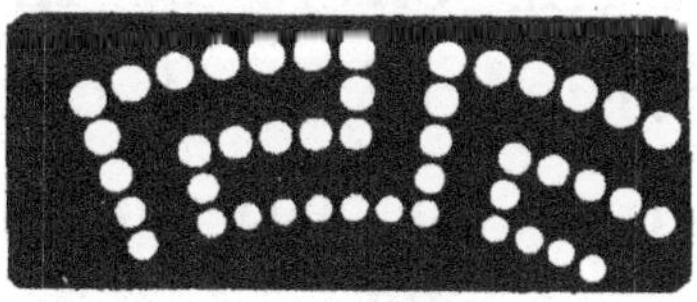

Eccentricity equal to 25¾.
Radius equal to ¾.

Decrease the eccentricity by turning the slide screw of the eccentric chuck forwards one turn and ten-sixteenths of a turn Decrease the radius by turning the slide rest screw forwards one-sixteenth of a turn; then proceed with the same angular tool as before. This set of circles completes the Etruscan Border, adapted to an eccentric circular division of ninety-six. It consists of as many circles as the exterior set, *viz.* eighty-four, which are to be described in the following manner: Set the circle of the eccentric chuck at division 96, and describe seven circles, one circle at that and at each succeeding division: miss one division, set the circle of the eccentric chuck at division 8, and describe seven circles as before: miss one division, set the circle of the eccentric chuck at division 16, and describe seven circles: and so on, till the set be completed at division 94.

SPECIMEN III.

This Specimen is composed of five sets of circles, of various radii, which are arranged round the common centre, at different eccentricities, in the manner represented in the accompanying Plate (III): by reference to this Plate, aided by the following data, the Turner will be enabled to produce a corresponding piece of Turning.

The tools to be used are double angular, Nos. 28 and 36; and the slide rest is to be set at right angles with the mandril of the lathe. The work to be turned (being first properly fixed in the eccentric chuck) must be brought to a plane and polished surface: then fix the angular tool No. 36, in the slide rest; and, after adjusting its point, the eccentric chuck, and the slide rest, to the common centre, with the greatest possible correctness, proceed to execute the centre, or interior set of circles.

FIRST SET OF CIRCLES.

Eccentricity equal to $1\frac{1}{4}$.
Radius equal to $1\frac{1}{8}$.

Produce the eccentricity by turning the slide screw of the eccentric chuck backwards one turn and a quarter. Produce the radius by turning the slide rest screw backwards one turn and an eighth of a turn. Then describe eight equidistant circles: these circles are to be cut sufficiently deep to make the figure, which is left by the intersection of the circles, sharp at top; and the same depth must be preserved throughout all the other sets.

SECOND SET OF CIRCLES.

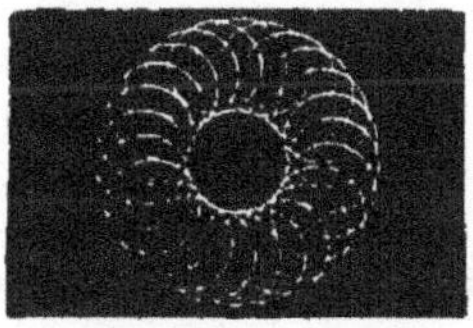

Eccentricity equal to $4\frac{1}{2}$.
Radius equal to $2\frac{1}{8}$.

To obtain the eccentricity turn the slide screw of the eccentric chuck backwards three turns and a quarter. To obtain the radius turn the slide rest screw backwards one turn. Then, with the same angular tool as before, describe twenty-four equidistant circles.

THIRD SET OF CIRCLES.

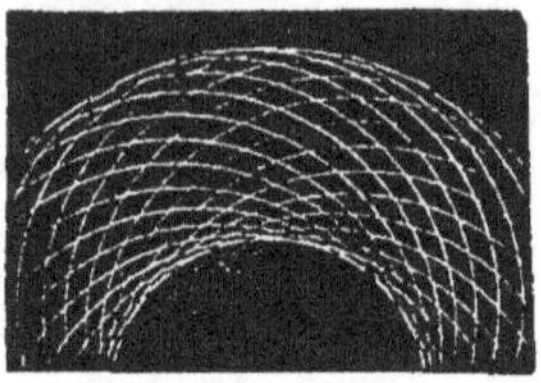

Eccentricity equal to $5\frac{1}{8}$
Radius equal to $11\frac{3}{4}$.

Produce the eccentricity by turning the slide screw of the eccentric chuck backwards five-eighths of a turn. Produce the radius by turning the slide rest screw backwards nine turns and five-eighths of a turn. Then, with the same tool, No. 36, as before, describe twenty-four equidistant circles.

FOURTH SET OF CIRCLES.

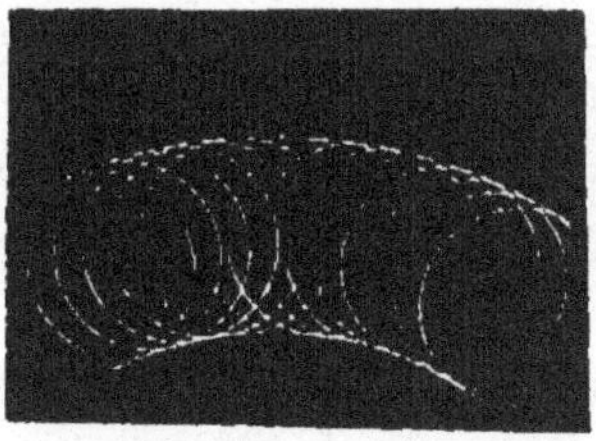

Eccentricity equal to $21\frac{1}{2}$.
Radius equal to $4\frac{3}{4}$.

Change the angular tool, and fix No. 28 in the slide rest; but, before it is fixed, adjust its point to the radius and depth of the circle last turned. To produce the eccentricity add to that required for the preceding set, by turning the slide screw of the eccentric chuck backwards sixteen turns and three-eighths of a turn. To produce the radius decrease that which was required for the preceding set, by turning the slide rest screw forwards seven turns. Describe ninety-six equidistant circles.

FIFTH SET OF CIRCLES.

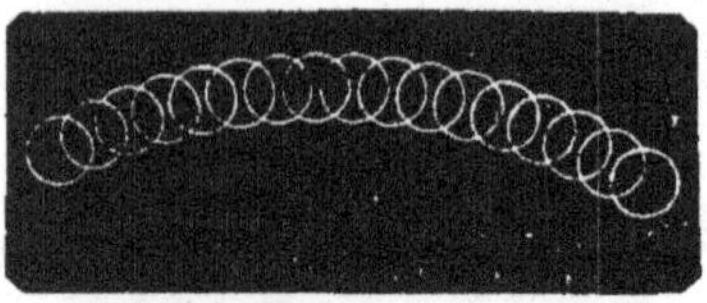

Eccentricity equal to $27\frac{3}{4}$.
Radius equal to $1\frac{1}{2}$.

Increase the eccentricity required for the set last described, by turning the slide screw of the eccentric chuck backwards six turns and a quarter; and decrease the radius by turning the slide rest screw forwards three turns and a quarter. Then describe ninety-six equidistant circles.

SPECIMEN IV.

This Specimen consists of five sets of circles, which form the outermost figures, and of five sets of arcs of circles, or curved lines, which form the centre figure. The circles are of different radii, and are arranged, at different eccentricities, round the common centre. The arcs of circles, or curved lines, are all of the same radius and eccentricity, and are produced from the point of the common centre, but vary in their measure. Plate IV represents the arrangement of these circles and arcs; which, together with the following practical directions, will enable the workman to execute a corresponding piece of Eccentric Turning.

In order to understand the description, hereafter given, of the method by which the arcs of circles, or curved lines, are to be produced, the following observations must be attended to.

It must be recollected, that, in Eccentric Turning, all circles and arcs of circles, are produced by the rotation of the lathe; that the radius of all circles, and of all arcs of circles, depend on the adjustment of the slide rest; and that the locality, or what is called eccentricity in this kind of Turning, of all circles, and of all arcs of circles, is assigned to them by the operation of the eccentric chuck. If a tool, therefore, be applied to the surface of the work, at any point without the centre of the lathe's motion, and the pulley wheel, which gives motion to the mandril of the lathe, be moved completely round, a circle will be described: if the pulley wheel be moved but half round, the tool will describe a semicircle, or an arc of 180 degrees: and if the pulley wheel be moved but a quarter round, the tool will describe a quadrant, or arc of 90 degrees. The pulley wheel which gives motion to the mandril of the lathe, affords, consequently, the practical means of measuring all arcs or curved lines; and it will give their measure both minutely and correctly, if a circle on the face of it be divided into some convenient number of parts. For instance, suppose such circle were divided into forty-eight parts, and that to produce an arc, or curved line, the said pulley wheel be moved forwards twelve of those parts, the measure of the arc, or curved line, thereby produced, would be equal to $\frac{12}{48}$; if the pulley wheel be moved twenty-four divisions, the measure of the curve would be equal to $\frac{24}{48}$, &c. &c. It is on this principle that the curved lines in question are produced, and that the measure of them is defined.

The work to be turned must be fixed in the eccentric chuck, and brought to a plane and polished surface; a double angular tool, No. 32, may be used throughout this piece of Eccentric Turning; and the slide rest is to be set at right angles with the mandril of the lathe.

Fix the angular tool in the slide rest, and adjust its point, the eccentric chuck, and the slide rest, to the common centre, with every possible degree of correctness. In this state of things, move the lathe round, and pressing the point of the angular tool gently against the surface of the work, describe a very minute dot; from which dot all the curved lines, which form the centre figure, are to be produced, in the following manner.

FIRST SET OF ARCS.

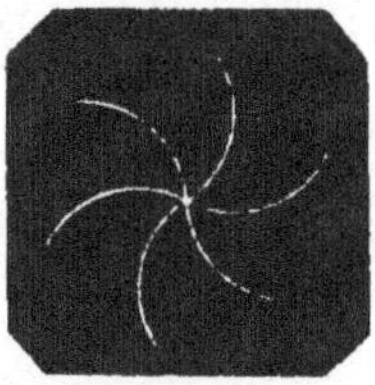

Eccentricity equal to 5.
Radius equal to 5.
Measure of the arcs $\frac{12}{48}$.

Produce the eccentricity and radius by turning the slide screw of the eccentric chuck and the slide rest screw backwards five turns each. Set the circle of the eccentric chuck at division 96, turn round the mandril pulley wheel until the dot, at the common centre, comes under the point of the tool; and then, setting the point of the tool in this dot, describe a curved line by moving the pulley wheel forwards (supposing it to be divided into 48) twelve divisions.[7] Then describe five other arcs, or curved lines, one at each

[7] It would tend very greatly to the correct execution of this centre figure, if stops were to be applied

of the following divisions of the circle of the eccentric chuck, 16, 32, 48, 64, 80.

to the pulley wheel, so as to secure it from moving, either way, beyond the intended limits of the curve.

Since the first edition was published I have seen various methods of stopping the pulley wheel, different to that I had, myself, adopted. Most of the lathe makers have now some contrivance they affix to their lathes, which they call the " segment stop." These contrivances are available for executing the arcs of circles in question, and all others; but why the lathe makers call it the segment stop I cannot say. It would have been better if it had been called otherwise, as there is no possible motion of the lathe that can describe the *segment of a circle*. Every curved line that is produced by the motion of the mandril of the lathe, must be either a circle, or an arc of a circle; whilst a segment of a circle is a superfices comprehended between two lines, *viz.* by a straight line denominated a chord, and by a curved line, denominated an arc.

SECOND SET OF ARCS.[8]

Eccentricity equal to 5
Radius equal to 5.
Measure of the arcs $\frac{11}{48}$.

These arcs, or curved lines, are to be produced from the centre dot, in the same manner as those of the former set, but the pulley wheel is to be moved only eleven divisions. There are twelve curved lines of this measure to be described, one at each of the under-named divisions of the circle of the eccentric ckuck, *viz.* 2, 14, 18, 30, 34, 46, 50, 62, 66, 78, 82, 94.

[8] In this and the following illustrations of the arcs, the lines which meet at the centre are those which are to be produced.

THIRD SET OF ARCS.

Eccentricity equal to 5.
Radius equal to 5.
Measure of the arcs $\frac{10}{48}$,

Produce six arcs, one at each of the following divisions of the circle of the eccentric chuck, *viz.* 4, 20, 36, 52, 68, 84.

FOURTH SET OF ARCS.

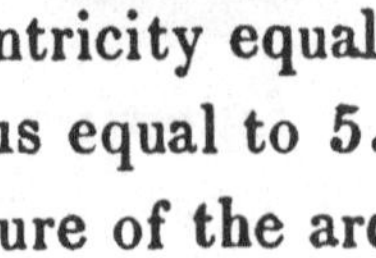

Eccentricity equal to 5.
Radius equal to 5.
Measure of the arcs $\frac{8}{48}$.

Produce six arcs, one at each of the following divisions of the circle of the eccentric chuck, *viz.* 6, 22, 38, 54, 70, 86.

FIFTH SET OF ARCS.

Eccentricity equal to 5.
Radius equal to 5.
Measure of the arcs $\frac{7}{48}$.

Produce twelve arcs, one at each of the following divisions of the circle of the eccentric chuck, *viz.* 8, 12, 24, 28, 40, 44, 56, 60, 72, 76, 88, 92.

FIRST SET OF CIRCLES.

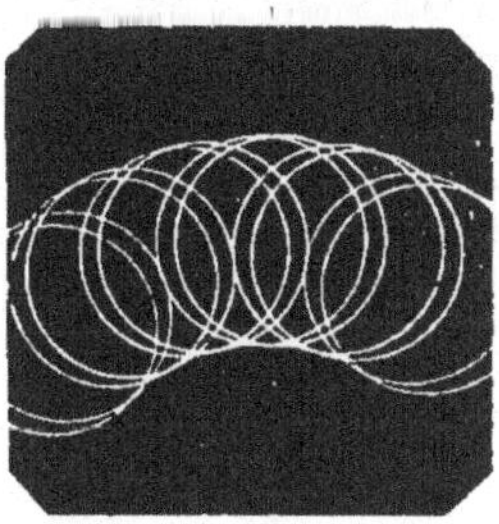

Eccentricity equal to $14\frac{1}{8}$.
Radius equal to 5.

Proceed with the same angular tool and radius as before. Increase the eccentricity by turning the slide screw of the eccentric chuck backwards nine turns and one-eighth of a turn; and then (supposing the circle of the eccentric chuck to be divided into 96) produce forty-eight circles in the following manner: *viz.*

Describe one circle at division 96: set the circle of the eccentric chuck at division 1, and describe another circle: miss two divisions, set the circle of the cecentric chuck at division 4, and describe a circle: set the circle of the eccentric chuck at division 5, and describe another circle; and so on, missing every alternate two divisions, till the forty-eight circles be completed.

SECOND SET CIRCLES.

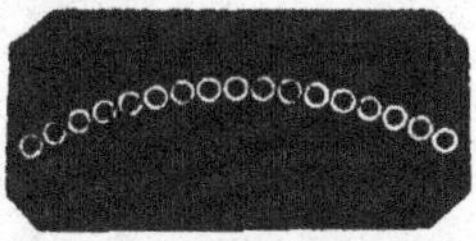

Eccentricity equal to $19\frac{3}{4}$.
Radius equal to $\frac{1}{2}$.

The same angular tool as before. Produce the eccentricity by turning the slide screw of the eccentric chuck backwards five turns and five-eighths of a turn. Produce the radius by turning the slide rest screw forwards four turns and a half. Describe ninety-six equidistant circles.

THIRD SET OF CIRCLES.

Eccentricity equal to $24\frac{1}{2}$.
Radius equal to 4.

The same angular tool as before. Produce the eccentricity by turning the slide screw of the eccentric chuck backwards four turns and three-fourths of a turn. Produce the radius by turning the slide rest screw forwards three turns and a half. Describe thirty-two equidistant circles.

FOURTH SET OF CIRCLES.

Eccentricity equal to $29\frac{1}{2}$.
Radius equal to $2\frac{1}{2}$.

The same angular tool as before. Increase the eccentricity by turning the slide screw of the eccentric chuck backwards five turns. Decrease the radius by turning the slide rest screw forwards one turn and a half. Describe thirty-two equidistant circles.

FIFTH SET CIRCLES.

Eccentricity equal to $30\frac{1}{2}$
Radius equal to 2.

The same angular tool as before. Increase the eccentricity by turning the slide screw of the eccentric chuck backwards one turn. Decrease the radius by turning the slide rest screw forwards half a turn. Describe thirty-two equidistant circles: these circles are not to be produced at the divisions of the circle of the eccentric chuck adopted in the third and fourth sets, but midway between those divisions.

SPECIMEN V.

THIS Specimen is composed of a number of concentric circles, and of arcs of circles, which cross each other in the manner represented in the accompanying Plate (V). The concentric circles increase in radius, in a uniform ratio from the common centre: the arcs of circles are all of the same measure, and are produced, not from the point of the common centre, but from points equidistant from it.

The tool to be used, to produce a piece of Eccentric Turning corresponding with this Specimen, is double angular No. 28, and the slide rest is to be set at right angles with the mandril of the lathe.

The work to be turned, being properly fixed in the eccentric chuck, is to be brought to a plane and polished surface. Then place an angular tool, No. 28, in the slide rest, adjust its point, the slide rest, and the eccentric chuck, to the common centre, and proceed to execute the concentric circles.

THE CONCENTRIC CIRCLES.

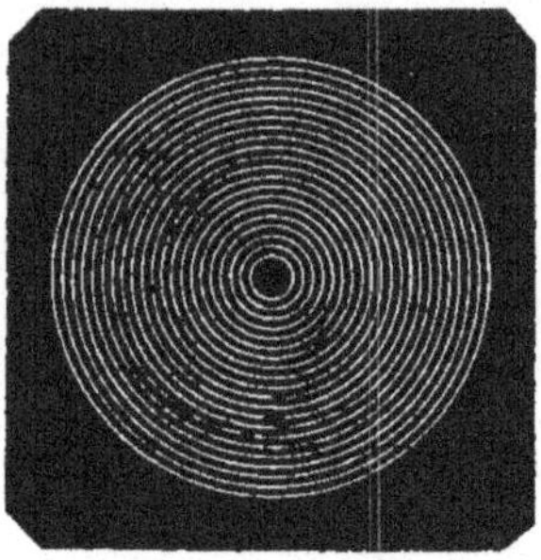

Radius of the first circle equal to 1.
Radius of the second circle equal to $1\frac{1}{2}$.
Radius of the third circle equal to 2, &c.

To produce the radius for the first, or internal circle, turn the slide rest screw backwards one turn. Describe one circle. To produce the radius for the second circle, increase the radius of the former by turning the slide rest screw backwards half a turn. Describe another circle; and so on, increasing the radius of each succeeding circle, by turning the slide rest screw backwards half a turn, until eighty circles are produced.

THE ARCS OF CIRCLES, OR CURVED LINES.

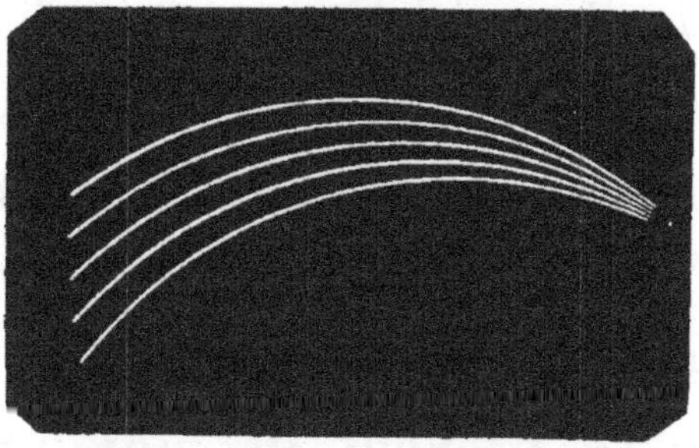

Eccentricity equal to 30.
Radius equal to 30.
Measure of the arcs[9] equal to $\frac{10}{48}$,

These arcs of circles, or curved lines, are to be cut with a similar angular tool,

[9] The method of producing and measuring the arcs, already described for Specimen IV, must be attended to in this Specimen. Vide p. 114.

No. 28, and to the same depth as the concentric circles. Supposing the radius of the exterior of the concentric circles to have been equal to $40\frac{1}{2}$, produce the radius for the arcs, or curved lines, and make it equal to 30, by turning the slide rest screw forwards ten turns and a half. Produce the eccentricity by turning the slide screw of the eccentric chuck backwards thirty turns. Turn the mandril pulley wheel, until the centre of the interior concentric circle coincides with the point of the tool. Move the pulley wheel forward, without letting the tool touch the work, equal to an arc of $\frac{11}{48}\frac{1}{2}$, and put on a stop to prevent its moving beyond that point. Move the pulley wheel back again equal to an arc of $\frac{10}{48}$, and put on a stop to prevent its moving any further that way.

The tool being now applied to the surface of the work, and the mandril pulley wheel moved through the space thus limited by the stops, an arc will be described equal to $\frac{10}{48}$; *viz.* an arc, which will begin at a distance from the common centre equal to $\frac{1\frac{1}{4}}{48}$, and end at a distance from the common centre equal to $\frac{11\frac{1}{4}}{48}$.

There are twelve sets of these arcs to be produced and arranged in the following manner; each set to consist of five arcs.

Set the circle of the eccentric chuck (supposing it to be divided into 96,) at division 96, and describe one arc at that and at the four succeeding divisions: miss three divisions, set the circle of the eecentric chuck at division 8, and describe five other arcs; and so on for the rest. The divisions of the circle of the eccentric chuck, at which the arcs in each set are to be described, are as under.

1st set of arcs, at Nos.	96,	1,	2,	3,	4.
2nd do.	8,	9,	10,	11,	12.
3rd do.	16,	17,	18,	19,	20.
4th do.	24,	25,	26,	27,	28.
5th do.	32,	33,	34,	35,	36.
6th do.	40,	41,	42,	43,	44.

7th set of arcs, at	48,	49,	50,	51,	52.
8th do.	56,	57,	58,	59,	60.
9th do.	64,	65,	66,	67,	68.
10th do.	72,	73,	74,	75,	76.
11th do.	80,	81,	82,	83,	84.
12th do.	88,	89,	90,	91,	92.

SPECIMEN VI.

The accompanying Plate (VI) represents the manner in which the sets of eccentric circles, and the sets of arcs, or curved lines, are to be arranged in this Specimen. The sets of circles are of different eccentricities and radii; and the arcs, or curved lines, are all of the same measure, and are produced from points without, and equidistant from, the common centre.

To form a piece of Eccentric Circular Turning, corresponding with this Specimen, use a double angular tool, No. 28, and set the slide rest at right angles with the mandril of the lathe.

The work to be turned is to be first properly fixed in the eccentric chuck, and brought to a plane and polished surface.

Then place the angular tool, No. 28, in the slide rest; adjust its point, the slide rest, and the eccentric chuck, to the common centre; and proceed to execute the arcs of circles, or curved lines.

THE ARCS, OR CURVED LINES.

Eccentricity equal to 12.
Radius equal to 12.
Measure of the arcs $\frac{7}{48}$.

Produce the eccentricity by turning the slide screw of the eccentric chuck backwards twelve turns. Produce the radius by turning the slide rest screw backwards twelve turns. Then adjust the mandril pulley wheel to the measure of the arcs of circles, or curved lines, in the same manner as directed for those in Specimen V.

These arcs of circles are, in all respects, similar to those described in Specimen V, except in their dimensions, and are to be described exactly in the same manner, at the same divisions of the circle of the eccentric chuck, at the same distance from the common centre, and should be cut sufficiently deep to bring the surface left between them to an angle at top.

CENTRE FIGURE, OR FIRST SET OF CIRCLES.

Eccentricity equal to $1\frac{1}{4}$.
Radius equal to 1.

Remove the stops from the mandril pulley wheel, and continue with the angular tool, No. 28. Produce the cecentricity by turning the slide screw of the eccentric chuck forwards ten turns and three-quarters of a turn. Produce the radius by turning the slide rest screw forwards eleven turns. Describe eight equidistant circles, of the same depth as the curved lines.

SECOND SET OF CIRCLES.

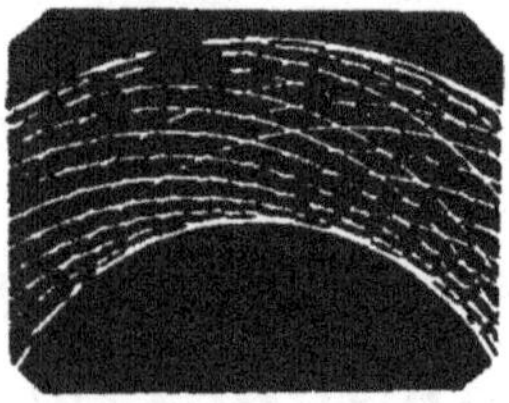

Eccentricity equal to $4\frac{1}{8}$.
Radius equal to 17.

The same angular tool, No. 28, as before, and the circles are to be cut to a sufficient depth to make the spaces, which their intersection leaves, sharp at top. Produce the eccentricity by turning the slide screw of the eccentric chuck backwards three turns and one-eighth of a turn. Produce the radius by turning the slide rest screw backwards sixteen turns. Describe twenty-four equidistant circles.

THIRD SET OF CIRCLES.

Eccentricity equal to 30.
Radius equal to $\frac{1}{2}$.

The circles in this and in the following seven scts, are to be produced by the angular tool, No. 28, and all cut to the same depth as the second set. Produce the eccentricity by turning the slide screw of the eccentric chuck backwards twenty-five turns and five-eighths of a turn. Produce the radius by turning the slide rest screw forwards sixteen turns and a half. Describe twelve equidistant circles.

FOURTH SET OF CIRCLES.

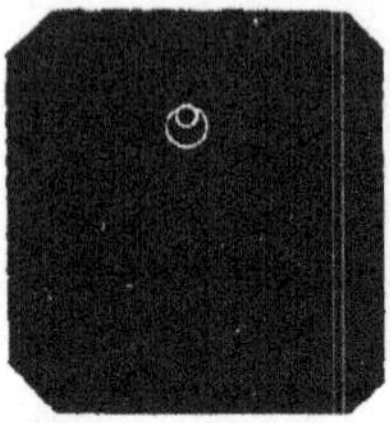

Eccentricity equal to $29\frac{1}{2}$.
Radius equal to 1.

The eccentricity is to be obtained by turning the slide screw of the eccentric chuck forwards half a turn. Obtain the radius by turning the slide rest screw backwards half a turn. Then describe twelve equidistant circles, which are to circumscribe those of the third set.

FIFTH SET OF CIRCLES.

Eccentricity equal to 29.
Radius equal to $1\frac{1}{2}$.

Produce the eccentricity by turning the slide screw of the eccentric chuck forwards half a turn. Produce the radius by turning the slide rest screw backwards half a turn. Describe twelve equidistant circles, which are to circumscribe those of the fourth set.

SIXTH SET OF CIRCLES.

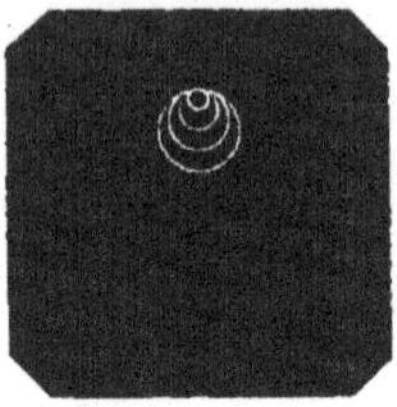

Eccentricity equal to $28\frac{1}{2}$.
Radins equal to 2.

The eccentricity is to be produced by turning the slide screw of the eccentric chuck forwards half a turn. Produce the radius by turning the slide rest screw backwards half a turn. Describe twelve equidistant circles, which are to circumscribe those of the fifth set.

SEVENTH SET OF CIRCLES.

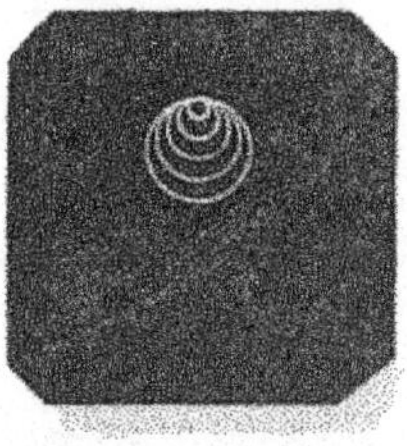

Eccentricity equal to 28.
Radius equal to $2\frac{1}{2}$.

Produce the eccentricity by turning the slide screw of the eccentric chuck forwards half a turn. Produce the radius by turning the slide rest screw backwards half a turn. Describe twelve equidistant circles, which are to circumscribe those of the sixth set.

EIGHTH SET OF CIRCLES.

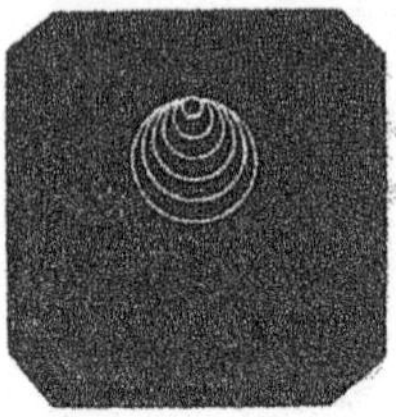

Eccentricity equal to $27\frac{1}{2}$.
Radius equal to 3.

Produce the eccentricity by turning the slide screw of the eccentric chuck forwards half a turn. Produce the radius by turning the slide rest screw backwards half a turn. Describe twelve equidistant circles, which are to circumscribe those of the seventh set.

NINTH SET OF CIRCLES.

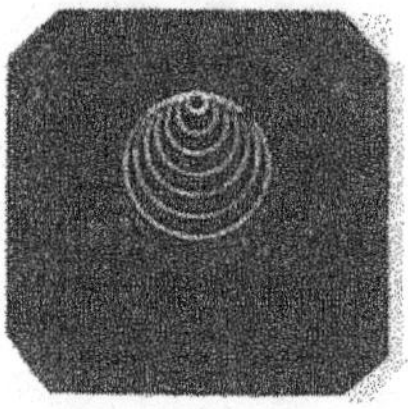

Eccentricity equal to 27.
Radius equal to $3\frac{1}{2}$.

Produce the eccentricity by turning the slide screw of the eccentric chuck forwards half a turn. The radius is to be produced by turning the slide rest screw backwards half a turn. Describe twelve equidistant circles, which are to circumscribe those of the eighth set.

TENTH SET OF CIRCLES.

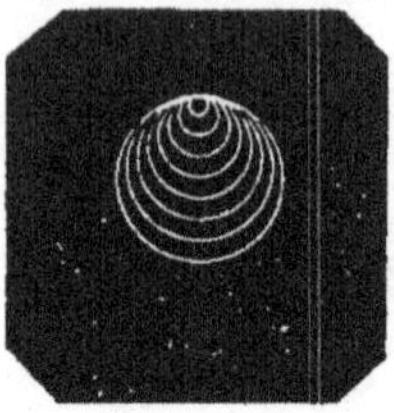

Eccentricity equal to $26\frac{1}{2}$.
Radius equal to 4.

To obtain the eccentricity turn the slide screw of the eccentric chuck backwards half a turn; and produce the radius by turning the slide rest screw forwards half a turn. Then describe twelve equidistant circles, which are to circumscribe those of the ninth set, and complete the figure.

ELEVENTH SET OF CIRCLES.

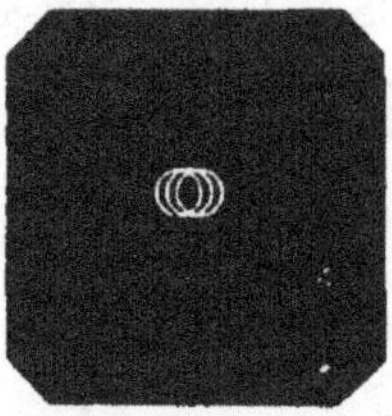

Eccentricity equal to 23½.
Radius equal to 1.

Proceed with the same angular tool, No. 28; and cut the circles sufficiently deep to make the spaces left between, or within the circles, sharp at top: this set of circles, as represented in the Plate (VI), is produced by a division of the circle of the eccentric chuck of 288.

Produce the eccentricity by turning the slide screw of the eccentric chuck forwards three turns. Produce the radius by turning the slide rest screw forwards three turns. Then set the circle of the eccentric chuck (supposing it to be divided into 288) at division 288: miss ten divisions, set the circle of the eccentric chuck at division 11, and describe one circle at that and at each of the two succeeding divisions, *viz.* at divisions 12 and 13: miss ten divisions, set the circle of the eccentric chuck at division 24, and describe one circle at that and at the two succeeding divisions, *viz.* at divisions 25 and 26; and so on, missing ten divisions, and describing three circles, until the figure represented in the Plate (VI) be completed.

The following may be substituted where the circle of the eccentric chuck is divided into only ninety-six.

ELEVENTH SET OF CIRCLES.

ADAPTED TO A DIVISION OF NINETY-SIX.

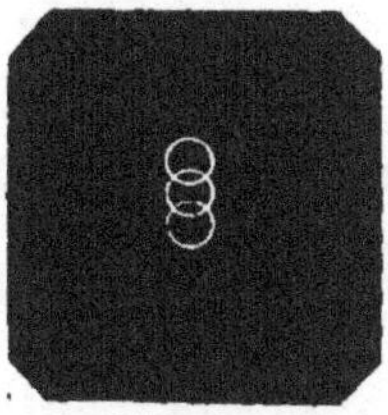

Eccentricity of the uppermost circle equal to 25.
Eccentricity of the centre circle equal to $23\frac{1}{2}$.
Eccentricity of the lowermost circle equal to 22.
Radius equal to 1.

Produce the eccentricity and radius, for the centre circle, in the manner directed for the set adapted to a division of 288, and describe one circle at each of the following divisions of the circle of the eccentric chuck, *viz.* at 4, 12, 20, 28, 36, 44, 52, 60, 68, 76, 84, 92.

Obtain the eccentricity of the uppermost circle by turning the slide screw of the eccentric chuck backwards one turn and half a turn; and, with the same radins, describe one circle at each of the divisions of the circle of the eccentric chuck, as before. Obtain the eccentricity for the lowermost circle, by turning the slide screw of the eccentric chuck forwards three turns; and, with the same radius as before, describe one circle at each of the divisions of the circle of the eccentric chuck before specified for the centre circle.

TWELFTH SET OF CIRCLES.

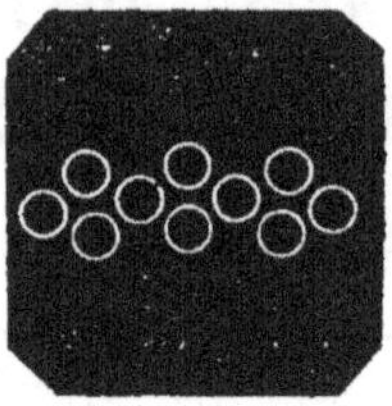

Eccentricity equal to 34.
Radius equal to 1.

Proceed with the same angular tool, No. 28, and cut the circles sufficiently deep to make their centres sharp at top: the lowermost of the three circles are those which are to be produced. Produce the eccentricity by turning the slide screw of the eccentric chuck backwards twelve turns. The radius remains as before. Describe forty-eight equidistant circles; one circle (supposing the circle of the eccentric chuck to be divided into 96) at each of the following divisions, *viz.* 2, 4, 6, 8, 10, 12, 14, 16, 18, 20, 22, 24, 26, 28, 30, 32, 34, 36, 38, 40, 42, 44, 46, 48, 50, 52, 54, 56, 58, 60, 62, 64, 66, 68, 70, 72, 74, 76, 78, 80, 82, 84, 86, 88, 90, 92, 94, 96.

THIRTEENTH SET OF CIRCLES.

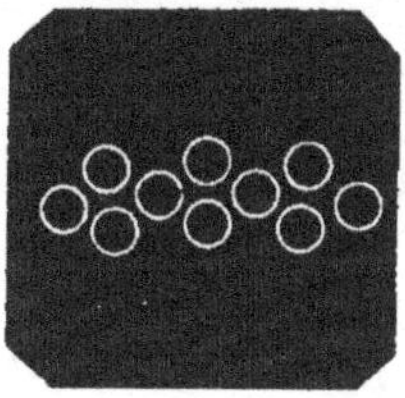

Eccentricity equal to 37.
Radius equal to 1.

The circles now to be described are the uppermost of the three circles, represented in this illustration. Produce the eccentricity by turning the slide screw of the eccentric chuck backwards three turns; and, with the same radius as before, describe one circle at each of the divisions of the eccentric chuck named for the twelfth set.

FOURTEENTH SET OF CIRCLES.

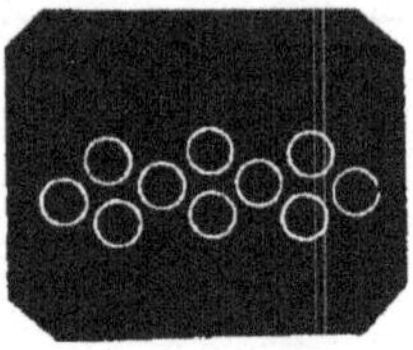

Eccentricity equal to $35\frac{1}{2}$.
Radius equal to 1.

The circles now to be produced are the centre of the three circles, represented in this illustration.

Produce the eccentricity by turning the slide screw of the eccentric chuck forwards one turn and a half; and, with the same radius as before, describe one circle (supposing the circle of the eccentric chuck to be divided into ninety-six) at each of the following divisions, *viz.* 1, 3, 5, 7, 9, 11, 13, 15, 17, 19, 21, 23, 25, 27, 29, 31, 33, 35, 37, 39, 41, 43, 45, 47, 49, 51, 53, 55, 57, 59, 61, 63, 65, 67, 69, 71, 73, 75, 77, 79, 81, 83, 85, 87, 89, 91, 93, 95.

The Specimens which follow, are illustrative of the powers of my COMPOUND Eccentric Chuck.

SPECIMEN VII.

This Specimen is intended to illustrate the power of the Chuck to produce squares, oblong figures, and all figures that result from the combination of two right line motions, acting, alternately, at right angles to each other. The piece of wood, or other material to be worked upon (in this case a piece of box wood was used), being firmly chucked, was faced down to a smooth surface and correct plane, with a tool fixed in the slide rest. The slide plates were placed to traverse at right angles, by adjusting the first circular movement (the one next the lathe head and between the two slide plates) for the index to point to division marked 3. The slide plate, next the lathe head, was

SPECIMEN VII.

then thrown out to *half* the extent of one side of the square required, by turning the slide plate screw *d. d*; and the second slide plate was thrown out, to an equal extent, by turning its screw the same number of turns. The describing point, or cutting tool, which had been previously fixed in the slide rest, will now point to one of the corners of the square, and must be adjusted to cut a circle of the diameter and depth required, and the lathe being moved round, the circle at the corner of the square was produced. The screw of the slide plate next the lathe head was then turned once round and another circle was cut; and so on, until twenty-six circles were cut, which completed one side of the square. The second slide plate was then worked, in the same manner as the other had been, and another side of the square was produced and made to consist, also, of twenty-six circles. The slide plate next the lathe head was then moved in the same manner as before, but in the contrary direction, and the third side of the square was obtained; and the other slide plate was then moved so as to form the fourth side and complete the square.

The internal figure, consisting of large circles, is effected on the same principle, by the action of the two slide plates.

It is obvious that right line patterns, similar to this Specimen and to others which are to be seen in the title page, cannot be obtained by any Chuck, *unless the slide plates throw out at both ends.*

M

SPECIMEN VIII.

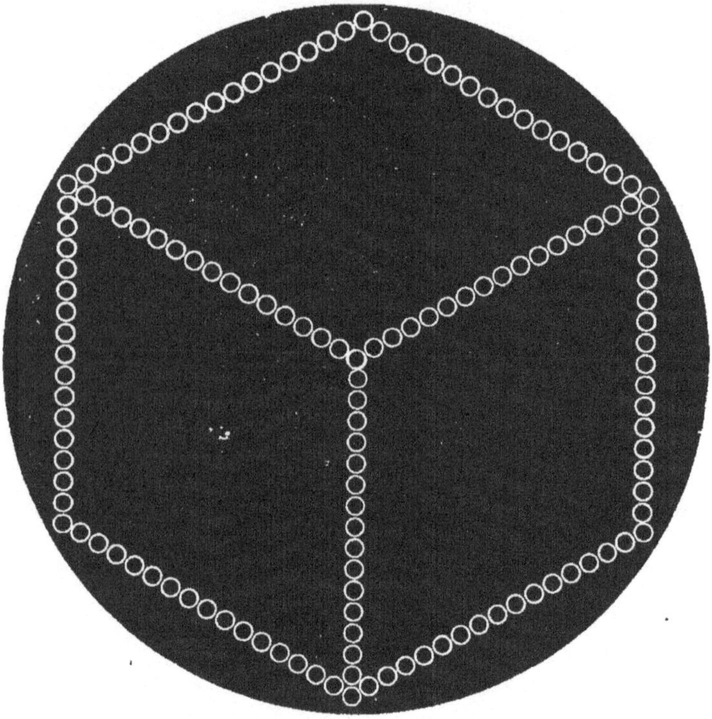

SPECIMEN VIII.

All figures similar to this Specimen require for their construction the whole of the adjusting powers of the Compound Chuck; and it will be found that it is not, altogether, immaterial in what order they are brought into action, to save labour and effect correct execution. But, without entering more particularly into the various modes of proceeding, I will merely detail, in the present instance, the method I pursued in executing the Engraving.

The piece of wood on which the engraving was to be made, having been firmly fixed in the chuck, was faced down by a tool in the slide rest, to a smooth surface and correct plane. The index to the circular movement next the lathe head was adjusted to division marked 8, and, thereby, the two slide plates (*i. e.* the two right line movements) were placed at an angle of 60 to each other. The index to the second circular movement (being that which is furthest from the lathe head) was adjusted to division marked 12. The slide plates were in their central position on their respective bed plates, and the index to each of them pointed, consequently, to division marked 0 of the scale on the edge of their bed plates.

An angular sharp-pointed tool was now fixed in the slide rest, and the point of it was adjusted (at the centre of the mandril's motion) to the radius of the circle required: the lathe was then turned round and the point of the tool described the circle which is seen in the centre of the engraving. The slide plate screw next the lathe head was now turned twice round to the left hand, thereby throwing out the slide plate, from its central position, to the extent of two divisions of the scale on the edge of its bed plate: the lathe was again turned round, and another circle was described by the tool next to the circle which had been made in the centre. The two other circles, next to the circle in the centre, were then described, by means of adjusting the second circular movement (*i. e.* that which is furthest from the lathe head) and causing its index to point to division marked 4, and then to division 8. The slide plate screw, next the lathe head, was again turned twice round, in the same direction as before, by which the slide plate became thrown out to the extent of *four* divisions of the scale on the edge of its bed plate; and, then, three more circles were described next to and in the same radial direction with the last, by causing the index to point to the same divisions as before, *viz.* divisions marked 12, 4, and 8. The slide plate was again thrown out to the extent of wo more divisions, and three more circles were produced next to those last described, and in the same radial lines as before; and so on, throwing out the slide plate to the extent of two turns of its screw, making three circles each time, and setting the circular movement for each circle, so that the index pointed to divisions marked 12, 4, and 8. This routine of adjusting the chuck was continued until each of the three radial lines of circles were made to consist of sixteen circles from the *centre* circle; and the slide plate screw next the lathe head was ***not again moved***—its index remaining stationary, pointing to the thirty-second division from the division marked 0 of the scale on the edge of the bed plate, during the execution of the remaining part of the engraving. The circular movement on the second bed plate was now adjusted for the index to point, in succession, to divisions marked 2, 6, and 10, and a circle was described at each of those divisions. The screw of the second slide plate (the one furthest from the lathe head) was now turned twice round, and the slide plate was thereby thrown out to the extent of two divisions from its central position, in an angular lineal direction to that in which the other slide plate had been worked; and six circles were described by setting the index of the circular movement to point to the divisions marked 2, 4, 6, 8, 10, and 12. The same slide plate was again thrown out to the extent of two divisions of the scale on the edge of its bed plate, and six more circles were described by setting the circular movement to the divisions marked 2, 4, 6, 8, 10, and 12, as before. The slide plate screw was again turned twice round, and six more circles were described at the same divisions as before; and so on until the engraving was completed.

SPECIMEN IX.

THIS Specimen is intended to exhibit the power possessed by a Compound Eccentric Chuck, of delineating or pointing out the path of an Ellipsis; and it will also serve as a practical illustration of what has been said at pages 16 and 17. Those who are conversant with the action of "Suardi's Geometric Pen" will very clearly see their way to the principle which assigns this property to an eccentric adjusting chuck; and will, as readily, trace it to the result of epicycloidal motion. It is well known that the periphery and curvature of every ellipsis are limited and defined by its diameters, and that the ellipsis will be more or less eccentric in proportion to the ratio in which its dia-

SPECIMEN IX.

meters differ in length. If they differ in the ratio of 100 to 1, that is, if the conjugate be only one-hundredth part as long as the transverse, the ellipsis will be extremely eccentric, and the curvature of its periphery will approach very closely to a right line: if they differ in the ratio of only 100 to 99, that is, if the conjugate be only one-hundredth part shorter than the transverse, then the ellipsis will approach very nearly to a circle.

In the case of the instrument before us, its power of adjustment for the attainment of the diameters of different ellipses is very great. The adjustment, only, of the slide plates to different eccentric positions, determine the diameters; and as the plates can be adjusted to every possible point of eccentricity, with regard to the centre of the mandril's motion and to each other, every variety of the ellipsis, from the right line to the circle, may be perfectly traced. The adjustments of the chuck may be differently combined, and brought into action under different arrangements, without causing any variation in the results; but, on the present occasion, I will confine myself to the method I pursued in working the chuck, to produce this Specimen, which, I think, will be found both ready and convenient; and if the following general rules be borne in mind and be strictly attended to, not any difficulty will be experienced in obtaining elliptical patterns of any eccentricity and size, that may be required, within the limit of the action of the chuck and lathe.

THE RULES.

1st. Elliptical patterns require the action of *both* the slide plates; and the *slide plate next the lathe head* is invariably thrown out to a *greater extent* than the other slide plate.

2nd. Both the circular movements are invariably brought into action, and the velocity of their adjustment must be in the ratio of 2 *to* 1 *in contrary direc-*

tions; that is, if the circular movement *next the lathe head* be moved 2 divisions at each adjustment, the other circular movement must be moved only 1 division at each adjustment, in the contrary direction. If the first be moved at the rate of 4 divisions, the second must be moved at the rate of 2 divisions; and so on.

3rd. The semi-transverse diameter is, invariably, equal to the extent of the two eccentric adjustments of the slide plates, added together. For example:

If the slide plate next the lathe head be adjusted to the extent of 20 turns of the screw, and the other slide plate to the extent of 10 turns, the semi-transverse will be equal to 30 turns of the screw.

4th. The semi-conjugate diameter is always equal to the difference between the extent of the two eccentric adjustments. For example:

If one slide plate be thrown out to

the extent of 20 turns of the screw, and the other slide plate to the extent of 10 turns of the screw, the semi-conjugate will be equal to 10 turns of the screw.

I will now explain, with reference to the foregoing rules, the adjustments of the chuck I adopted in executing that portion of the Specimen IX which consists of a set of forty-eight consecutive white lines surrounding black dots, through the centre of which the path of a true ellipsis lies.[10] At page 17 I explained the manner of tracing an ellipsis, the diameters of which were to be in the ratio of 3 to 2;

[10] All curves traced on the principle here described will be true, in every point, to that particular section of the Cone which is known and distinguished by English mathematicians, as the *Ellipsis of Apollonius*; and there is no method, by means of points, by which that section of the Cone can be more easily and correctly delineated. Figures, also, of a true *oval* shape may be traced by means of this Chuck, as perfectly as by any method heretofore suggested by the mathematician; and in exemplifi-

and in the case of the Specimen before us the diameters are in the ratio of 5 to 3.

The piece of wood on which the engraving was executed, having been firmly fixed in the chuck, the face of it was turned over by a tool in the slide rest, and brought to a smooth surface and true plane. The slide plate next the lathe head (rule 1) was thrown out of its central position and adjusted by turning its screw round 24 turns; and the other slide plate was adjusted by turning its screw 6 turns, and throwing the slide plate out in the same lineal direction in which the other slide plate had been thrown out. Now 6 added

cation of its possessing this power, I have executed with it and introduced two Specimens, No. XIII, and No. XIV.

By different combinations, arrangements, and adjustments of the Chuck, other patterns may be produced in endless variety; and, in proof of it, the engravings in the title page will serve as a few examples. I will also give a few more; but, with regard to these and the oval figures, it is not my intention to enter into any detail of the adjustments of the Chuck by which they were produced.

to 24 equals 30 (3rd rule), and thereby the semi-transverse became equal to 30 turns of the screw, and the whole transverse to 60 turns. The difference between the two adjustments (24 and 6) is 18, and therefore (4th rule) the semi-conjugate was equal to 18 and the whole conjugate to 36 turns of the slide plate screw: thus the slide plates of the chuck were adjusted to cause the length of the diameters to be in the ratio of 60 to 36, or as 5 to 3.

The index of each of the circular movements pointed to division marked 12, and a pointed tool being fixed in the slide rest and adjusted to the radius of the circle required, the lathe was turned round and a circle was described at one of the extreme ends of the transverse diameter. The circular movement, *next the lathe head* (2nd rule), was advanced 4 divisions, and the other circular movement was advanced 2 divisions, *in the contrary direction,* and another circle was described. The circular movement next the lathe head was again advanced 4 divisions, and the other circular movement was advanced two divisions, as before, and another circle was described; and so on until the whole forty-eight circles were produced, and the Specimen, so far, finished.

The diameters of the ellipsis mentioned at page 17 were in the ratio of 3 to 2. The diameters of the Specimen I have just described are in the ratio of 5 to 3; and I will now give the adjustments of the slide plates of the chuck, by which other ellipses may be obtained of various diameters.

The eccentric adjustment of the Slide Plates.		The length of the Diameters of the Ellipses.		The ratio of difference between the length of the Diameters.	
Slide Plate next the Lathe Head.	The Second Slide Plate.	Transverse Diameter.	Conjugate Diameter.	Transverse.	Conjugate.
No. of turns of the screw.	No. of turns of the screw.	No. of turns of the screw.	No. of turns of the screw.		
54	6	60	48	5 to	4
33	3	36	30	6 —	5
36	6	42	30	7 —	5
69	3	72	66	12 —	11
58$\frac{1}{2}$	1$\frac{1}{2}$	60	57	20 —	19
298$\frac{1}{2}$	1$\frac{1}{2}$	300	297	100 —	99
151$\frac{1}{2}$	148$\frac{1}{2}$	300	3	100 —	1

SPECIMEN X.

A line passed through the centres of the circles which form this Specimen, will trace the path of a true Epicycloidal curve; and one of the names which have been assigned to this curve is "The Snail of Pascal." The following arrangement of the chuck will produce it.

The piece of wood, on which the engraving was executed, having been firmly fixed in the chuck, the face of it was turned down to a true plane and smooth surface; and a pointed tool was fixed in the slide rest and was adjusted to the radins of the circles, 96 of which form the pattern. The circular movements were adjusted to cause each index to point to the division marked 12. The slide plate

SPECIMEN X.

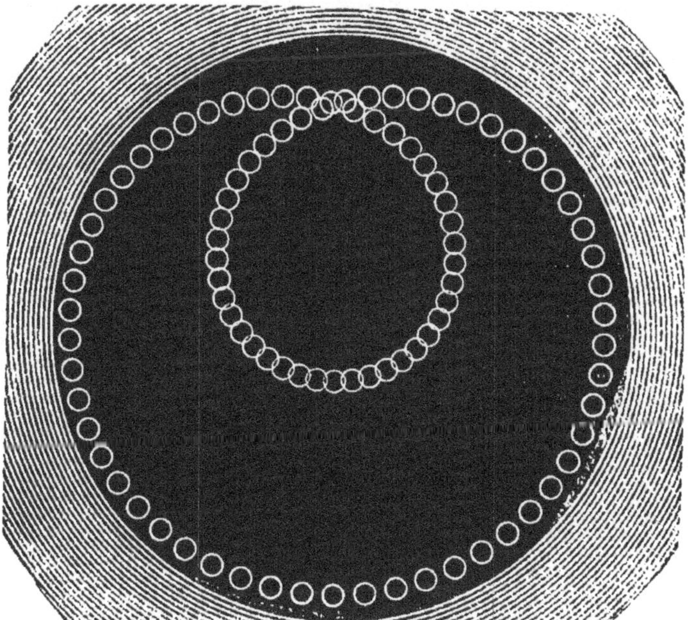

next the lathe head was thrown out to the extent of sixteen turns of its screw. The other slide plate was thrown out, in the same direction, to the extent of twelve turns of its screw. A circle was then described. The circular movements were both moved, in the same direction, one division (equal to $\frac{1}{96}$ part of the whole circle) and another circle was described; and so on, moving each circle one division, in the same direction, and describing a circle each time, until the figure was completed.

SPECIMEN XI.

A line passed through the centres of the circles which form this Specimen will trace the path of a true Epicycloidal curve; and this curve comprehends the curve of M. Carré. The arrangements of the chuck and slide rest to produce it, are as under.

The piece of wood, on which the engraving was executed, was firmly fixed in the chuck, and the face of it was turned down to a true plane and smooth surface. A pointed tool was then fixed in the slide rest and adjusted to the radius of the *largest* of the forty-eight circles which form the pattern. The circular movements were adjusted for the index of each to point to the division marked 12. The slide plate

SPECIMEN XI.

next the lathe head was thrown out to the extent of eight turns of its screw. The other slide plate was thrown out, in the same direction, to the extent of sixteen turns of its screw; and a circle was then described. The circular movements were both moved in the same direction, two divisions, (equal to $\frac{2}{96}$ parts of the whole circle) and the radius of the circle was decreased, by turning the slide rest screw one-twelfth part round, and then another circle was described. The circular movements were again moved two divisions each, in the same direction; and the radius of the circle was further decreased by turning the slide rest screw another twelfth part round, and another circle was described. And so on, moving the circles two divisions, and decreasing the radius of the circle each time by turning the slide rest screw one-twelfth of a turn, until the index to each circle pointed to division marked 6, when the figure will have progressed to the smallest of the 48 circles; and, then, the radius of the remaining circles is to be increased, at each adjustment of the circular movements, by turning the slide rest screw one-twelfth part of a turn, in the contrary direction, until the index to each circular movement again pointed to division marked 12, when the pattern was completed.

SPECIMEN XII.

SPECIMEN XII.

A line passed through the centres of the circles which form this Specimen will trace the path of a true Epicycloidal curve; and the arrangements of the chuck, for producing the outermost set of circles, are as under.

The piece of wood on which the engraving was executed, was firmly fixed in the chuck, and the face of it was turned down to a true plane and smooth surface. A pointed tool was then fixed in the slide rest and adjusted to the radius of the small circles, ninety-six of which form the pattern. The circular movements were adjusted for the index, to each of them, to point to division marked 12[illegible]. The slide plate next the lathe head was thrown out to the extent of twenty-two turns of its screw; and the other slide plate was thrown out to the extent of five turns of its screw. A circle was then described. The circular movement next the lathe head was then moved two divisions; and the other circular movement was moved one division, in the same direction; and then another circle was described: and so on, moving the circular movement next the lathe head two divisions and the other circular movement one division, and describing a circle each time, until the whole ninety-six circles were completed.

The internal curve is of the same character and proportionate to the other, and is composed of the same number of circles of larger diameter.

CONCLUSION.

THE Specimens which I have now given will have developped, in some degree, the system by which various geometrical figures may be delineated by combining a right line with a circular adjustment; and it is quite obvious that the path of every possible curve and figure may be mathematically traced by means of such an instrument as a Compound Eccentric Chuck. The Geometric Pen, invented by The Count Suardi, is capable of delineating an infinite variety of curves, by the continuous motion of a point acting against a fixed surface. My Compound Eccentric Chuck is capable of delineating and recording the path of the same curves, by means of dots and circles; and the Specimens I have given will shew, though to a very limited extent, that it possesses this power. I will only add, that it will afford me pleasure to explain my Chuck to any Gentleman, who is an Amateur Turner, and who may be desirous of adding to his Turning Apparatus such an instrument as my Compound Eccentric Chuck. Any application addressed to me shall be immediately attended to, and I will appoint a time when I will be at home to shew the Chuck.

J. H. Ibbetson.

3, Brompton Grove,
Knightsbridge.

SUPPLEMENTARY SPECIMENS

Referred to in the Note at page 165.

SPECIMEN XIII.

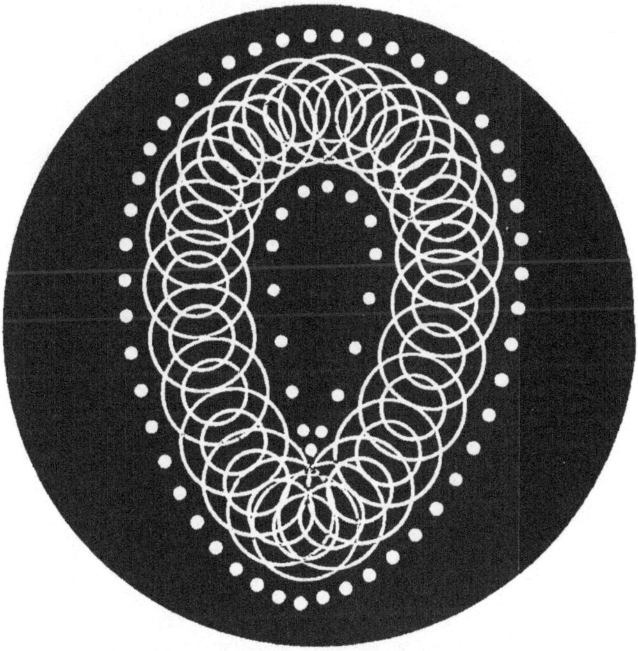

SPECIMEN XIV.

Spotted something that isn't quite right?

Let us know and we'll fix it for future reprints. And as a thank you for your diligence, we'll send you a free Sothis book of your choice.

publisher@sothis.press

This page intentionally left blank

Except of course for that line above. And this one.
And whilst we have your attention, perhaps you'd
like to follow us on Instagram @sothis.press. Go on.

Made in the USA
Monee, IL
08 November 2026

67978720R00121